Guidebook to the Geology of Northern and Western Idaho and Surrounding Area

Guidebook to the Geology of Northern and Western Idaho and Surrounding Area

Edited by
V. E. Chamberlain
Roy M. Breckenridge
and
Bill Bonnichsen

Idaho Geological Survey
University of Idaho
Moscow, Idaho 83843

Bulletin 28
1989

ISBN 1-55765-027-6
First Printing, May 1989

CONTENTS

CHAPTER TWO
Idaho Batholith and Accreted Terranes

CHAPTER THREE
Tectonic and Sedimentary Sequences in Northeast Idaho and Northwest Montana

CREDITS FOR COVER PHOTOGRAPHS

FRONT COVER: top, unnamed glacial horn in granite of the Idaho batholith, north part of the Frank Church River of No Return Wilderness, Roy M. Breckenridge; bottom left, Precambrian Belt rocks along the Kootenai River at Moyie Springs, Boundary County, Idaho, Roy M. Breckenridge; bottom center, Bunker Hill glory hole (1897-1904), Coeur d'Alene Silver District, Idaho, University of Idaho Barnard Stockbridge collection; and bottom right, Pleistocene Bonneville Flood gravels and White Butte, a tuff cone near Walters Ferry along the Snake River, Idaho, Bill Bonnichsen.

BACK COVER: left, altered rocks in the western border zone of the Idaho batholith, South Fork drainage of the Salmon River, Roy M. Breckenridge; and right, trilobites from a Cambrian concretion, collected near Heron, Montana, Elizabeth A. Measures.

PREFACE

This guidebook is the third Bulletin recently produced by the Idaho Geological Survey that presents current geologic information covering broad areas of Idaho. Bulletin 26, *Cenozoic Geology of Idaho*, describes in forty-two articles the state's geologic history for the last 65 million years. Bulletin 27, *Guidebook to the Geology of Central and Southern Idaho*, contains twenty-one articles written on that extensive region of the state. The book was prepared from field trip articles for the Geological Society of America's Rocky Mountain Section meeting held in 1988 at Sun Valley.

The present guidebook, Bulletin 28 – *Guidebook to the Geology of Northern and Western Idaho*, is organized around field trips for the May 1989 joint meeting in Spokane, Washington, of the 42nd Annual Rocky Mountain Section and the 85th Annual Cordilleran Section of the Geological Society of America. The book consists of nine field trips in Idaho and adjoining parts of Oregon and Montana. A companion volume, produced for the same joint meeting by the Washington Division of Geology and Earth Resources, covers field trips in Washington and adjacent parts of Canada.

These professional meetings in 1988 and 1989 have provided the Idaho Geological Survey with a singular opportunity to bring together in two trip guides a sampling of the diverse geology of Idaho. The books also answer the increasing demand from the geologically interested public for more detailed information about Idaho's land and earth resources.

The publication of these two volumes is largely the result of the Idaho Geological Survey moving forward into the era of desktop publishing. For the 1989 book, the Survey used WordPerfect® 5.0 word processing software and Xerox Ventura Publisher® 2.0 design and composition software on a Cactus® AT/12 computer interfaced with a Hewlett Packard LaserJet Series II® printer. The computer hardware and software made it possible for the Survey's two-person editorial staff to produce an attractive publication in time for the meeting.

We acknowledge the effort of many individuals responsible for this volume. Roger C. Stewart directed the production and assisted in the technical editing. Jennifer Pattison

Hall designed the book and prepared the camera-ready layout. External reviewers for each of the articles were chosen and acknowledged by the authors themselves; and we thank all those who assisted in this review process. Photomechanical line art was prepared by Printing and Design Services and the Department of Geography's Carto-Graphics Laboratory at the University of Idaho. Daryle Faircloth drafted the route map for the *Introduction to the Guidebook*.

V. E. Chamberlain
Roy M. Breckenridge
Bill Bonnichsen
Moscow, Idaho
May 1989

INTRODUCTION TO THE GUIDEBOOK

V. E. Chamberlain[1]
Roy M. Breckenridge[2]
Bill Bonnichsen[2]

The geologic history of northern and western Idaho and surrounding areas can be summarized very roughly into four main episodes. During the Precambrian and the Paleozoic, sedimentary deposition occurred on a passive continental margin. This was followed by a succession of collisional orogenies during the Mesozoic, as terranes were added to the western edge of the Paleozoic continent. The Cenozoic was characterized by crustal extension and volcanism, which continue today, and the Quaternary was marked by a sequence of glaciations. The nine field trips in this guidebook are divided among four chapters that cover different aspects of the geologic history. The first three chapters are arranged chronologically, and the fourth concerns the world famous Coeur d'Alene Mining District. The routes of all the field trips are shown on Figure 1, where they are identified by their authors.

The first chapter, *Late Cenozoic Lake Environments*, contains three field trips. The first trip, *Pleistocene Ice Dams and Glacial Lake Missoula Floods in Northern Idaho and Adjacent Areas* (see Figure 1, No. 1), examines the flood deposits of Glacial Lake Missoula, a Pleistocene

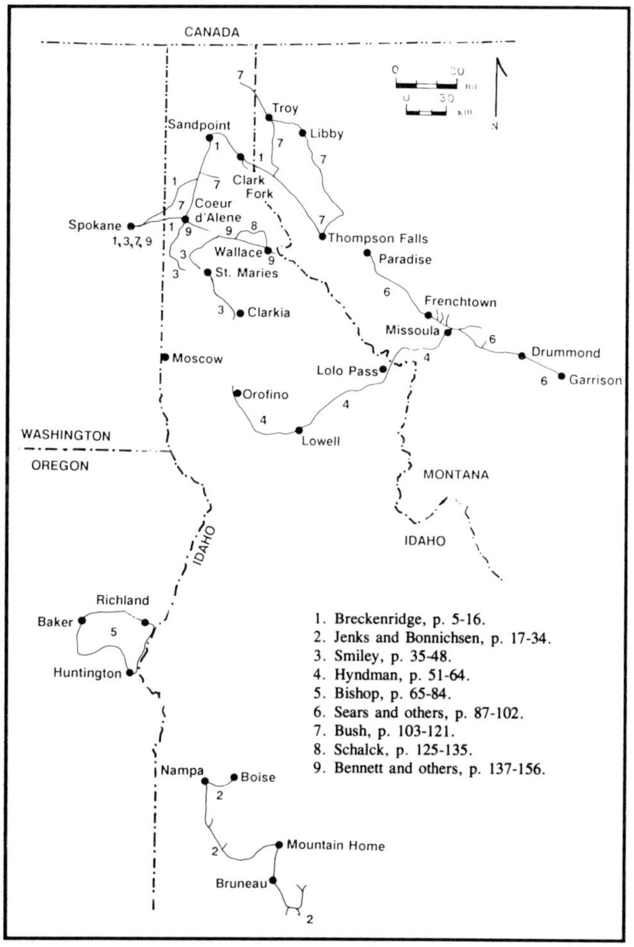

Figure 1. Index map to field trips.

1 Department of Geology and Geological Engineering, University of Idaho, Moscow, Idaho 83843.
2 Idaho Geological Survey, University of Idaho, Moscow, Idaho 83843.

ice-dammed lake that covered 3,000 square miles in western Montana. Periodic breaching of the ice dam near Clark Fork, Idaho, gave rise to multiple catastrophic flood episodes that are the largest floods known on the earth. These floods traveled westward, following the present-day Pend Oreille, Spokane, and Little Spokane River systems, and scoured the Channeled Scabland of Washington. The field trip studies the geology of the ice dam area and the character of the flood deposits between the ice dam and Spokane. The second field trip, *Subaqueous Basalt Eruptions Into Pliocene Lake Idaho, Snake River Plain, Idaho* (see Figure 1, No. 2), visits the basin of Lake Idaho, the name given to a sequence of ancient lakes which inundated a deepening graben in the western Snake River Plain and adjacent areas during the Pliocene. The largest of these lakes reached the size of present Lake Ontario. Lake Idaho drained as the Snake River carved out Hells Canyon along the Idaho-Oregon border. Subsequent smaller lakes formed as the Snake River became temporarily dammed by various and numerous local basalt lava flows. The two-day field trip surveys the evidence for the high stand of Lake Idaho as well as the spectacular volcanic products formed by the interaction of lavas and lake waters. The third field trip in this chapter, *The Miocene Clarkia Fossil Area of Northern Idaho* (see Figure 1, No. 3), examines another lake formed as a result of damming by lava flows, in this case by Miocene Columbia River basalts blocking the St. Maries River. The one-day outing to Clarkia concentrates on the unique collection of flora and fauna that accumulated in the bottom sediments and swamps of Miocene Clarkia Lake. These remnants of a time past have been amazingly well-preserved over the last 16 million years. In some specimens, the original color and organic chemistry can still be detected.

Chapter Two, *Idaho Batholith and Accreted Terranes,* looks at some of the results of Mesozoic collisional orogenies, in two separate two-day trips. The fourth field trip of the guidebook, *Formation of the Northern Idaho Batholith and the Related Mylonite of the Western Idaho Suture Zone* (see Figure 1, No. 4), crosses a well-exposed portion of the northern Idaho batholith. The trip reviews the sequence of events that led to the formation of the batholith and the surrounding high-grade metamorphic rocks. This field trip also visits the Kamiah plutonic complex and the western Idaho suture zone, along which cordilleran accreted terranes were annealed to the North American craton. The fifth field trip, *Transect Through the Baker and Wallowa-Seven Devils Terranes, Northeastern Oregon* (see Figure 1, No. 5), makes a detailed examination of the geology of two accreted terranes.

Chapter Three, *Tectonic and Sedimentary Sequences in Northeast Idaho and Northwest Montana*, covers a Paleozoic sedimentary sequence and a Mesozoic thrust plate. The sixth field trip, *A Structural Section Through a 25-Km-Thick Thrust Plate in West-Central Montana* (see Figure 1, No. 6), studies a cross-section through a thrust system on the southwestern limb of the Purcell anticlinorium in west-central Montana. The trip traverses the transition from ductile deformation in the biotite grade of metamorphism, near Paradise, right through to the brittle behavior of the thrust sheet in the foreland basin, near Garrison. The seventh field trip, *The Cambrian System of Northern Idaho and Northwestern Montana* (see Figure 1, No. 7), involves a two-day detailed study of Middle and Upper Cambrian sedimentary cycles. The sediments were deposited originally in the miogeocline of a continental margin and in water depths varying from peritidal to below wave base.

Chapter Four, *Coeur d'Alene Mining District*, brings together the geology and mining history of one of the nation's important mineral areas. The eighth field trip, *The Geology and Alteration of the Gem Stocks, Shoshone County, Idaho* (see Figure 1, No. 8), is a one-day in depth examination of the Gem stocks, a highly differentiated group of Tertiary alkaline intrusive rocks. The magmatic phases, mineralization, and internal features of the stocks are examined along with the alteration and effect on the enclosing Precambrian metasediments of the Belt Supergroup. The last trip of the guidebook, *The Geology and History of the Coeur d'Alene Mining District, Idaho* (see Figure 1, No. 9), explores in one day both the history and the geology of this mining district that claims the greatest recorded silver production in the world. The underground tour was made by special arrangement for registrants at the Geological Society of America's May 1989 Meeting in Spokane, Washington, and may not be available to the casual visitor.

We hope that visitors and local residents alike will enjoy using this guidebook. We trust that the explanations given in the nine field trip articles will enhance their appreciation of the geologic framework that underlies the splendid scenery to be found in northern and western Idaho.

Chapter One
Late Cenozoic Lake Environments

The St. Joe River, the famous "river in a lake," is constrained by natural levees through Lake Chatcolet near St. Maries, Idaho. The lake basin is formed in Miocene basalt flows. *Photograph courtesy of Idaho Division of Travel Promotion.*

Pleistocene Ice Dams and Glacial Lake Missoula Floods in Northern Idaho and Adjacent Areas

Roy M. Breckenridge [1]

INTRODUCTION

In 1923 Professor J Harlen Bretz of the University of Chicago began a series of research papers explaining the origin of the Channeled Scabland in Washington. He attributed this system of dry channels, coulees, and falls to an episode of flooding on a scale larger than geologists had ever recognized. His "outrageous hypothesis" was disputed by prominent geologists, and the resulting controversy is one of the most famous in geologic literature. Bretz's ideas for such large-scale flooding challenged the uniformitarian principles then ruling the science of geology. Bretz nevertheless persisted in his thinking and, after additional evidence showed a source for the flood (Pardee, 1942), found his ideas finally accepted.

The source of the immense flood was attributed to the sudden drainage of glacial Lake Missoula. A great lobe of the Cordilleran ice sheet had advanced down the Purcell Trench and dammed the Clark Fork Valley. The resulting glacial lake held over 2,000 km^3 of water. The failure of the ice dam, the cause of the flood, released as much as 62.5 km^3/hour (Baker, 1973) through the Clark Fork valley.

This field trip guide describes the Late Pleistocene geology of the ice dam area near Clark Fork, Idaho, and the area initially inundated by outburst flooding from glacial Lake Missoula. The guide begins at Spokane, Washington, and proceeds to Coeur d'Alene, Idaho, via Interstate 90 (I-90) then north on U.S. Highway 95 (U.S. 95) to Sandpoint, east on Idaho State Highway 200 (S.H. 200) to the town of Clark Fork, and from there by paved county road to Heron, Montana. An optional return route

EDITOR'S NOTE: This article appeared in slightly different form as: *Evidence for the ice dams and floods in the Purcell Trench* published in Washington Division of Geology and Earth Resources Information Circular 86, *Geologic Guidebook to Washington and Adjacent Areas,* edited by N. A. Joseph and others, 1989.

[1] *Idaho Geological Survey, University of Idaho, Moscow, Idaho 83843.*

Breckenridge, R. M., 1989, Pleistocene ice dams and glacial Lake Missoula floods in northern Idaho and adjacent areas, *in* V. E. Chamberlain, R. M. Breckenridge, and Bill Bonnichsen, editors, Guidebook to the Geology of Northern and Western Idaho and Surrounding Area: Idaho Geological Survey Bulletin 28, p. 5-16.

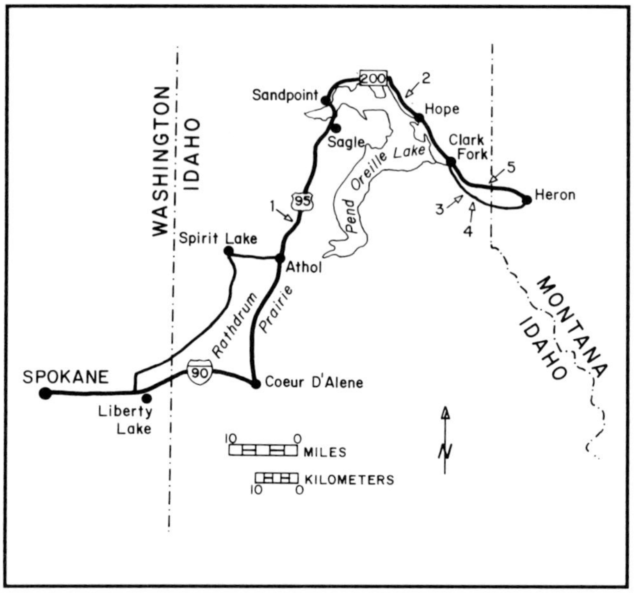

Figure 1. Route map of field trip and location of stops.

diverges at Athol, Idaho, and returns to Spokane by Spirit Lake and Rathdrum, Idaho. Figure 1 shows the trip route and the highways and roads traveled.

The locations of geologic features and stops are keyed to the highway mileage markers along the roadside. On the Clark Fork-Heron loop where there are no markers, the mileage is given.

DISCUSSION

Cordilleran Ice Sheet

During Pleistocene time the Cordilleran ice sheet advanced southward from Canada into Montana, Idaho, and Washington (Figure 2). A number of large glacially dammed lakes including glacial Lake Missoula formed along its margins. The regional extent of the Cordilleran ice sheet has been summarized by Waitt and Thorson (1983) and more recently by Richmond (1986). However, the area of the ice dam itself has received little attention. Unpublished mapping by Breckenridge and by Kiver and Stradling for the *Quaternary Geologic Atlas of the United States* (U.S. Geological Survey, in prep.) has refined the ice boundaries in northern Idaho and northeastern Washington (see discussion in Richmond, 1986). Figure 2 shows the mapped extent of the ice in the Idaho area and the location of the ice dam and outburst routes.

Evidence of pre-late Wisconsin glaciation in the Idaho panhandle is sparse, probably due to obliteration by later glaciation and flooding. However, older till buried beneath the latest till is exposed by cuts in the Purcell

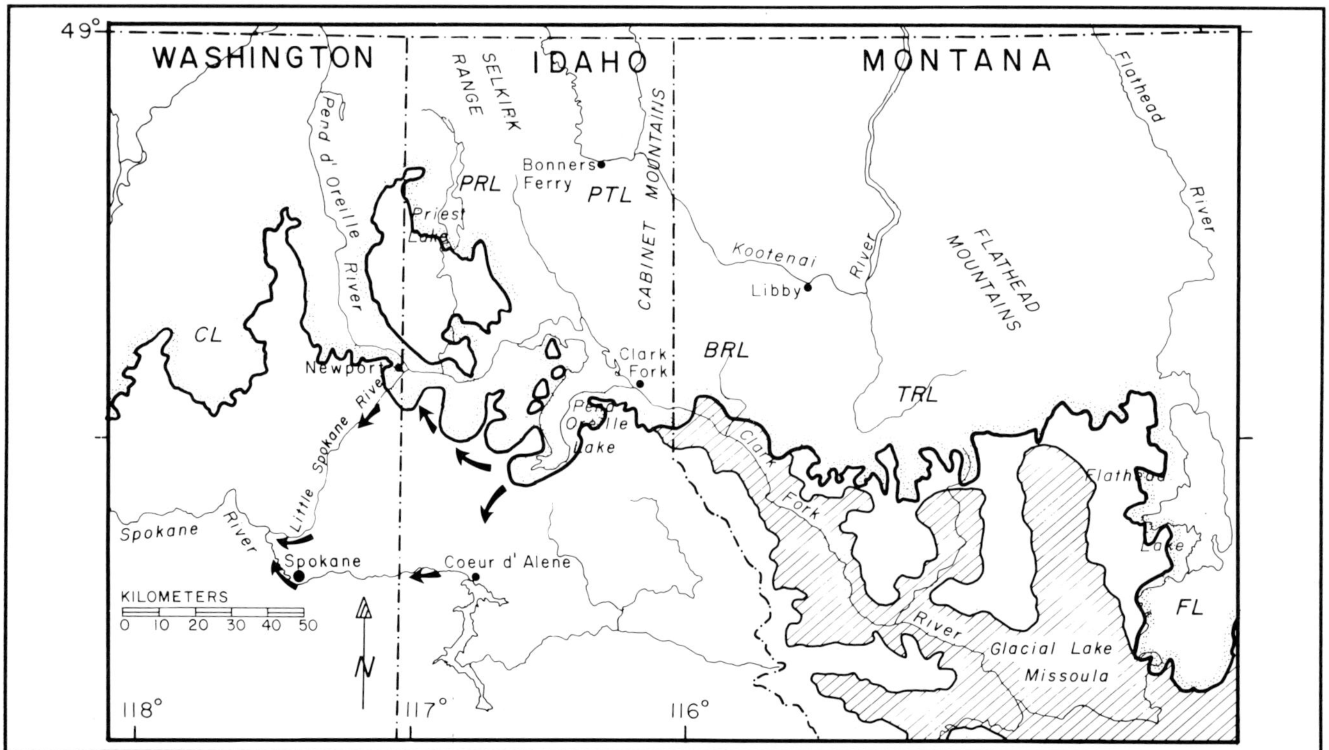

Figure 2. Map of Cordilleran ice margin (modified after Richmond, 1986), glacial Lake Missoula, and major flood outburst routes. BRL = Bull River Lobe; FL = Flathead Lobe; PTL = Purcell Trench Lobe; PRL = Priest River Lobe; TRL = Thompson River Lobe. Glacial Lake Missoula = cross-hatched area. Arrows show major flood outburst routes.

Trench near Cocolalla Lake and at Moravia north of Sandpoint. This older till and the glacial record in other parts of the Rocky Mountains indicate that an early advance or advances of the Cordilleran ice sheet covered nearly all of the Selkirk and Cabinet mountains (Alden, 1953; Richmond, 1986; and Breckenridge, unpublished map). Richmond (1986) placed the latest advance of the Pend Oreille lobe of the Cordilleran ice sheet between 25 and 13 Ka with an interstadial represented by the 17 Ka ice-free date of Clague (1978, 1981). Waitt (1985) interpreted the same evidence to show that glacial Lake Missoula could have existed only between about 17 and 11 Ka.

Source for the Flood

J. T. Pardee first reported on studies of glacial Lake Missoula in 1910, but not until 1942 did he present new evidence for rapid drainage of the large lake dammed by a lobe of the ice sheet (Pardee, 1910, 1942). His discussion of the unusual currents necessary to produce flood features in the lake basin provided the long-awaited source for the flooding that Bretz (1923) proposed to have formed the Channeled Scabland. Alden (1953) in his summary of the glacial geology of western Montana and adjacent areas of Idaho discussed the aspects of the ice dam and the multiple glacial advances, and he speculated on the lake and drainage routes of glacial Lake Missoula. Chambers (1971) described the sedimentation cycles of rhythmically bedded bottom sediments and interpreted them as multiple episodes of emptying and filling.

Pardee (1942) calculated that glacial Lake Missoula had a volume of 2,167 km^3 when it stood at an elevation of 1,265 meters against the ice dam and estimated that the lake drained at a rate of 15.3 km^3/hr. Baker (1973) estimated the maximum rate of discharge to be 62.5 km^3/hr, and new calculations by Craig and Hanson (1985) set the lake volume at that elevation to be 2,514 km^3. These huge discharge magnitudes and discharge rates are responsible for the megaripples and other large-scale sedimentary and geomorphic features formed by the floods.

Character of the Ice Dam and Location of Outburst Routes

Chambers (1971) noted the unstable characteristics of glacial dams and discussed four proposed mechanisms for their failure: (1) subglacial failure by flotation (Thorarinsson, 1939), (2) deformation of ice by water pressure (Glenn, 1954), (3) enlargement of subglacial tunnels by water (Liestøl, 1955), and (4) enlargement of tunnels by icebergs (Aitkenhead, 1960). Waitt (1985) proposed jökulhlaup releases when the depth of ice-impounded water at a dam reaches 90 percent of the thickness of the ice. He suggested that the increasing hydrostatic pressure progressively wedges into the ice-rock seal until the ice becomes buoyant. Flowing water then forms and enlarges subglacial tunnels. Finally the tunnel roof collapses and the ice dam fails causing the lake to empty. Waitt used the self-emptying model to explain the numerous rythmite sequences preserved in areas flooded but adjacent to main courses of flood. He interpreted each cycle in these rhythmite sequences as a separate flood. Recently Craig (1987) modeled the interaction of water and ice dams and simulated cycles of ice damming, lake filling, and erosion of the ice dam using global climatic variations. Costa (1988) compared the jökulhlaup models with modern ice-dammed lakes and regarded the data as too diverse and the extrapolations too great to be reliable.

The main path of the Lake Missoula floods (Figure 2) was from the end of Pend Oreille Lake through the Rathdrum Prairie and Spokane Valley. Another major flood route followed the Pend Oreille River, crossed divides, and followed the Little Spokane River valley. Other ice-dammed lakes existed in the valleys proximal to the Cordilleran ice sheet, and flooding from those sources must have also coursed down the Columbia River system. Another huge Pleistocene flood, from pluvial Lake Bonneville in Utah, traveled down the Snake River and the lower Columbia River sometime between 14 and 15 Ka (Scott and others, 1982). Both the Bonneville and Missoula flood deposits occur in the Lewiston basin and have been described by Webster and others (1982).

ROAD LOG

Spokane to Coeur d'Alene

From Spokane proceed east on I-90.

The flood gravels of the Spokane Valley and Rathdrum Prairie are the sole-source aquifer for the greater Spokane area (see Molenaar, 1988). Numerous wells tap this popularly termed "underground river" in the Pleistocene flood gravels.

Milepost Description

289 Pines Road. North of I-90 a large Missoula flood pendant bar trails to the west (lee side) from behind the scoured bedrock knobs. Note the prominent terrace of Missoula flood gravels south of the freeway.

296 Liberty Lake-Otis Orchards exit 296. South of I-90 note the scoured bedrock and the terraced gravel bars that dam the Liberty Lake basin. These deposits were interpreted by some to be moraines (Weis and Richmond,

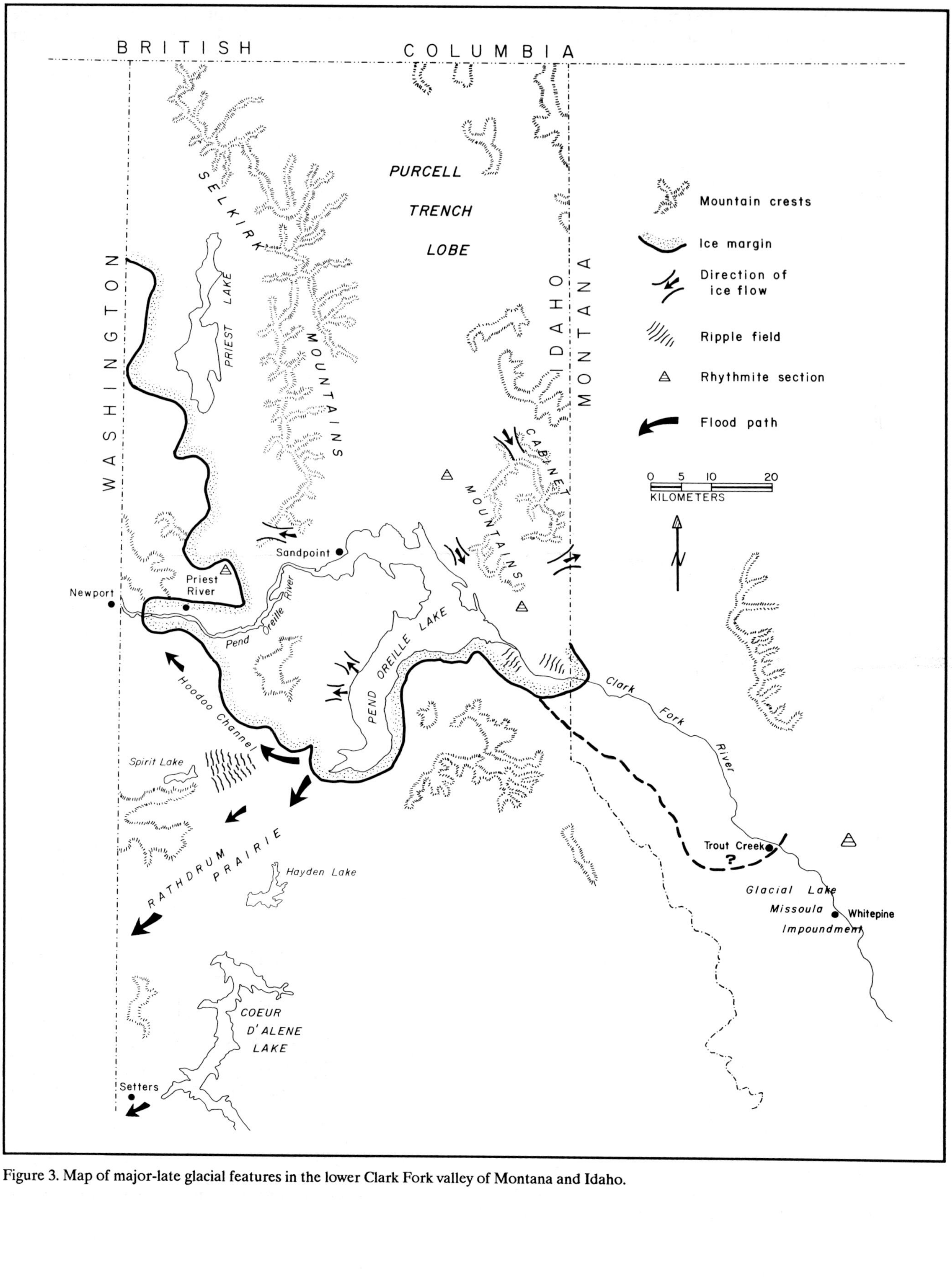

Figure 3. Map of major-late glacial features in the lower Clark Fork valley of Montana and Idaho.

1965). McKiness (1988) mapped the south side of the Spokane Valley and the Rathdrum Prairie and found no conclusive evidence of glaciation.

298 Missoula flood bar on south side of I-90.

0 Washington-Idaho state line, Spokane River, and Port of Entry. Newcomb (1953) interpreted a seismic cross section of the valley and estimated the base of the gravel aquifer to be at an elevation of 488 meters near the state line, as compared to a base level of 567 meters on the basalt at the Spokane Falls. This apparent deepening of the valley toward the east may be due to scour by ice and floods or to erosion by preglacial streams (see discussion at Stop 2).

2 Pleasant View Road. Jacklin Seed Farm on south. The ground water of the Rathdrum Prairie and the Spokane Valley is used to irrigate one of the world's largest producing areas of bluegrass lawn seed.

5 Post Falls. Here in 1891 Frederick Post purchased a waterfall site for a saw and grist mill from Chief Seltice. By 1906 Washington Water Power Company (WWP) had tied their dams at Post Falls and Spokane Falls to a 60,000 volt line that served the Coeur d'Alene mining district (Conley, 1982). The Post Falls facility now produces 15,000 kilowatts and controls the level of Coeur d'Alene Lake and the flow of the Spokane River.

8 Huetter Rest Area.

12 Junction I-90 and U.S. 95 in Coeur d'Alene. Follow exit signs to U.S. 95 north toward Sandpoint. Coeur d'Alene has a history rooted in mining, logging, and recreation, but actually the city began as a military fort established by General William Tecumseh Sherman in 1879. Before good roads were built, lake steamboats carried passengers, towed logs, and hauled freight and silver ore (Conley, 1982). Coeur d'Alene Lake (second largest in Idaho) is dammed by Lake Missoula flood gravels and is the source of the Spokane River, a tributary of the Columbia River. Flood waters of glacial Lake Missoula inundated the area currently occupied by the lake basin and at times overtopped a pass to the south at Setters (Dort, 1960; Figure 3). Some workers have put the ice advance as far south as Coeur d'Alene (Flint, 1936; Alden, 1953; Dort, 1960). Glacial grooves (?) have been reported on bedrock near Post Falls and at Tubbs Hill in Coeur d'Alene. Weathered deposits just to the north of Coeur d'Alene near Hayden Lake have been interpreted as old till (Richmond, 1965) or Tertiary sediments (Conners, 1976). Figure 3 shows the late-glacial ice and flood features in northern Idaho.

Coeur d'Alene to Sandpoint

Proceed north on U.S. 95.

Milepost Description

447 Columbia River basalt cliffs on the east. Missoula floods scoured basalt blocks from the margins of the Rathdrum Prairie and Spokane Valley. Large basalt blocks and boulders are not uncommon in the Spokane aquifer (C. M. Breckenridge, Dickerson Pump and Irrigation Co., oral communication, 1988).

450 Town of Athol. U.S. 95 crosses the large gravel plain that splays from the southern end of Pend Oreille Lake (Figure 4). The plain is composed of till near the end of the lake, but here within several kilometers from the lake, it grades into flood gravels with large-scale ripples. Rathdrum Prairie has no surface drainage, but large volumes of ground water flow through the gravels.

452 Hoodoo Creek Rest Area. The Hoodoo valley is an abandoned channelway that discharged flood and meltwater from the terminus of the Pend

Figure 4. Aerial view of the south end of Pend Oreille Lake. Glacially scoured cliffs in background. Foreground is mostly till molded by Missoula floods. Town of Bayview on left. A state park now occupies the site of the Farragut Naval Training Station, dismantled after WWII. (Post-WWII photograph courtesy of Idaho Travel Committee.)

Oreille Lake sublobe. Current direction indicators show that at least the latest phases of flooding traveled northwest through the channel toward Newport, Washington.

453 **Stop 1: Flood gravels.**

Turn left on Granite Lake road just past the rest area. Proceed to railroad crossing and stop at the gravel pit. Here crudely bedded flood gravels dip northward along the Hoodoo channelway. Note the scoured bedrock on the railroad right-of-way just to the west across the road from the gravel pit. **Caution! Be alert to fast approaching trains.** Return to U.S. 95 and proceed north.

456 Careywood. A well-developed moraine marks the southernmost recognized extent of the latest Wisconsin ice advance in this valley.

463 Cocolalla Lake. A moraine dammed lake. A pre late-Wisconsin till is exposed in roadcuts on the right. The ice limit on Huckleberry Mountain to the west is at least 1,160 meters.

464 Westmond. The ice limit on Little Blacktail Mountain, the highest mountain to the east, is as high as 1,220 meters. Ice from the Pend Oreille Lake sublobe moved west through several passes on Little Blacktail Mountain.

467 Algoma Lake, Heath Lake. Core from Algoma Lake to the west of the highway contained Mazama ash (6.7 Ka) at a depth of 1.2 meters and a yet unidentified ash near 2.7 meters. Note the scoured bedrock at the valley margin and elongated hills of till in the valley.

469 The Sagle area is composed mostly of till rather than flood deposits. Logs of wells here show blue clay and till. Ground water is relatively scarce, and septic drainage is poor due to impervious clay in these glacial deposits (DeSmet, 1983). This part of the valley apparently was not subjected to flood scour in latest-glacial time. Either the late-glacial flooding was not extreme or part of the ice lobe remained intact here while most of the Missoula flood water passed down the Pend Oreille Lake basin.

470 Gravel pits on west. Interbedded Missoula flood sands and gravels dip steeply to the south. Here the Pend Oreille River valley was a flood channel. Flood deposits are common to the west (Figure 3).

472 U.S. 95 crosses the north arm of Pend Oreille Lake. The first bridge built here was completed in 1910. It was nearly 2 miles long and was reported to be the world's longest wooden bridge.

Sandpoint to Clark Fork

Follow U.S. 95 signs through Sandpoint. Just north of Sandpoint at the junction of U.S. 95 and Idaho State Highway (S.H.) 200 proceed straight ahead on S.H. 200 toward Clark Fork.

Our route now crosses the Purcell Trench, along which a main lobe of the Cordilleran ice sheet moved south from Canada (Figure 3). The ice sheet was joined by alpine glaciers from the Selkirk Range on the west as well as by ice from the Cabinet Mountains on the east. The Schweitzer Basin ski area is visible above Sandpoint on the west. The route along the north side of Pend Oreille Lake passes through some excellent examples of crag and tail topography—glacially scoured knobs and molded till.

Milepost Description

38 Cross the Pack River, a major drainage of the southern Purcell Trench and the Selkirk Range.

44 **Stop 2: "Lake Missoula" highway geologic sign.**

Vista of Pend Oreille Lake. The lake, over 350 meters deep, is the largest in Idaho and one of the largest in North America. The lake basin is impounded by moraines at the south end. Bathymetry of the lake shows immense deltas from the Clark Fork and the Pack Rivers. Like most of the world's large deep lakes, Pend Oreille is the subject of numerous legends, including bottomless depths and monsters like the "Pend Oreille Paddler" (McLeod, 1987).

Origin of the lake basin has been debated. Its anomalous depth compared to adjacent ice-covered areas in the trench may be due to a preglacial river valley, more erodible sediments, or flood scour (Anderson, 1927; Savage, 1965; Conners, 1976). An interpretation of the lake basin and sediment based on new seismic reflection data is shown in Figure 5. These data are similar to seismic information from deep glacial lakes in Europe studied by Finckh and others (1984). The bottom of the lake basin is U-shaped and shows ice scour. Some evidence of fault control on the east side of the basin is shown by the seismic profile. Although the basin was scoured earlier, it was not scoured or flushed of sediments by the latest episodes of flooding but rather it contains over 75 meters of glaciolacustrine sediments. Samples collected by

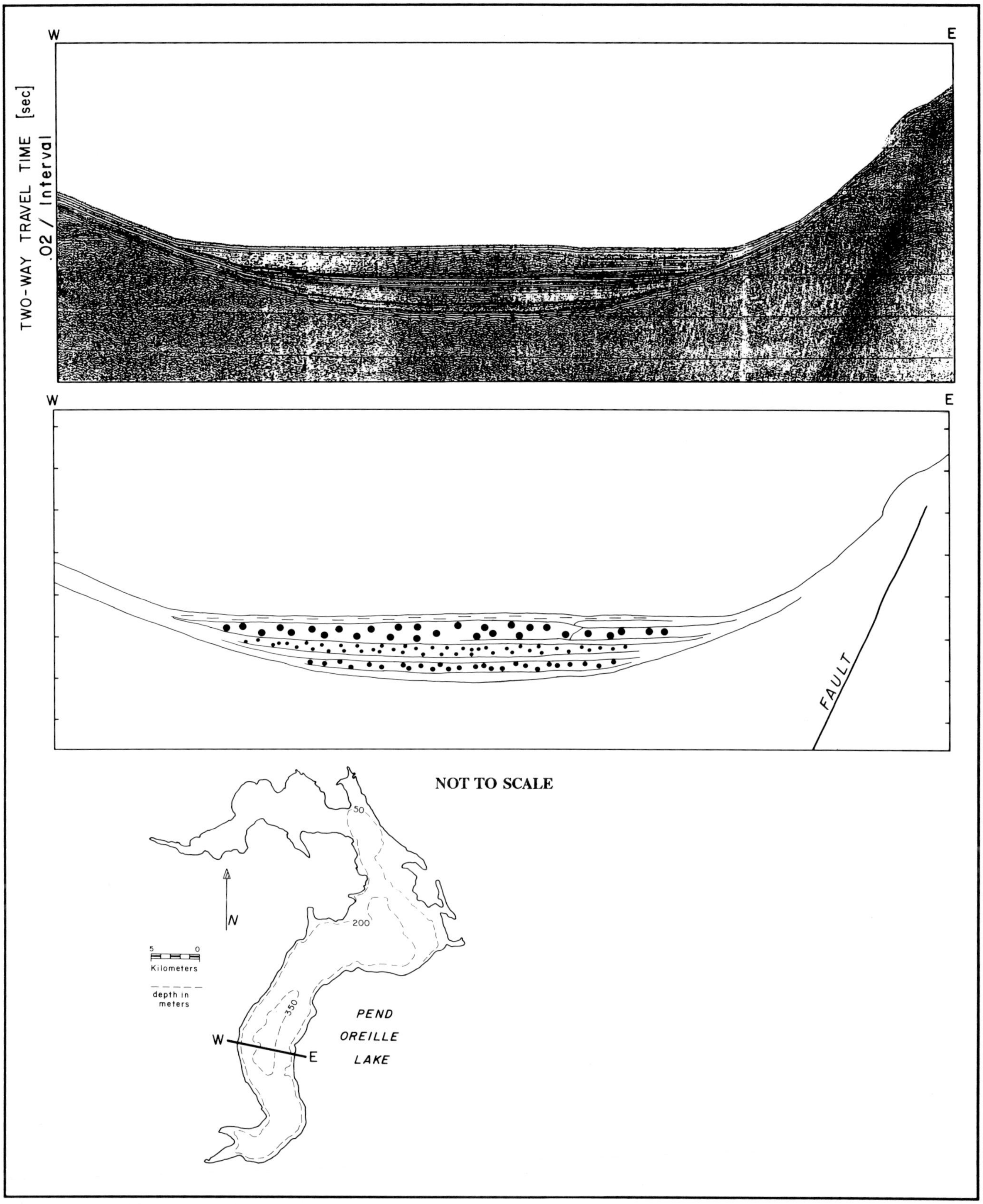

Figure 5. Map, seismic reflection profile, and interpretive reflection profile across southern part of Pend Oreille Lake (based on data provided by U.S. Navy, David Taylor Research Facility, Bayview, Idaho).

the U.S. Navy indicate the uppermost layer is Holocene ooze. At least two and probably four major depositional units exist below the uppermost layer. Although at present subsurface sampling cannot substantiate the content of the lower units, these units are probably coarse flood or till deposits based on comparison with other lakes (Finckh and others, 1984).

45 Bridge over boat basin.

49 Samowen Road. Kullyspel House historical sign at junction on right. The Samowen peninsula exhibits ice-scoured bedrock and gravel deposits.

50 Denton Slough.

51 Delta of Clark Fork River to the south.

55 Lightning Creek Bridge. Lightning Creek is clogged with coarse Pleistocene flood and glacial boulders moved from upvalley terraces and redeposited here by periodic flash floods. Enter town of Clark Fork.

Clark Fork-Heron Loop

The route through Clark Fork was an important one in the early exploration and settlement of the region. The first white explorer known to have visited the area was David Thompson, an Englishman who traveled through this area in 1809. He established fur trading posts at Selish House near Thompson Falls and at Kullyspell House near Hope. Both trading posts were short-lived because the Indians of the area – the Flathead, Kalispel, Spokan and Coeur d'Alene Tribes – were not traditional trappers and the marauding Blackfeet were hostile toward traders (Conley, 1982). David Thompson is best known, however, as the pioneer geographer and explorer during that era in northwest America. The natives called him "koo koo sint," the man who looked at stars. In 1876, the Northern Pacific Railway chose the Clark Fork route, and soon towns like Thompson Falls and Hope sprang up as sidings. These towns were the local jumping off spots to the lumber and mining camps. Many prospectors were bound for the Coeur d'Alene district, just 40 kilometers to the south, and now the largest silver-producing district in the world.

The Clark Fork-Heron loop is not marked by mileposts; therefore, mileage is given for the next two stops.

Mileage Description

0.0 In Clark Fork turn south on marked road to Lakeview. Use caution crossing railroad tracks at south end of town. Cross bridge over the Clark Fork River and turn east (left).

All water draining from glacial Lake Missoula passed through the mouth of the Clark Fork valley. At peak flow as much as 2.1 x 10^7 m^3/sec of water from Lake Missoula passed through the valley. Note the water gap on the south skyline at an elevation of 1000 meters. The road crosses a large ripple field (Figure 6) formed by flood waters below the Castle Rock cataract. A thin blanket of lacustrine silt overlies the flood gravels indicating that the later stages of Lake Missoula drained quiescently. The ripple field is the site of the Ruen seed potato farm, the first and only such farm in northern Idaho.

6.5 **Stop 3: Flood bar stratigraphy.**

At approximately 6.5 miles the road ascends terrace gravels. Stop at the freshest exposure in the roadcut. The road is excavated in a flood bar. The quality of the exposures is variable because of slumping. The bedding sets are several meters thick, and foresets dip in a downstream direction

Figure 6. View from Castle Rock to the west at the mouth of the Clark Fork valley. All flood water from glacial Lake Missoula passed through the field of view. Note the water gap on the south (left) skyline at 1000 meters. In the center of the view, long wave length ripples are marked by the tree lines through the fields.

Figure 7. Photograph of coarse flood deposits exposed in gravel pit. Bedding dips westward down the Clark Fork valley. Section of gravel freshly exposed by slump is 8 meters high.

up to 20 degrees. The clasts are Precambrian Belt Supergroup rocks and are subrounded to rounded. Individual beds are generally only poorly sorted. A perpendicular face in a private gravel pit (Figure 7) just to the north under the powerlines exposes well-developed scour and fill troughs. The gravels are overlain by about a meter of silt. The section appears to have no major breaks in deposition. Note the texture, lithology, and bedding of these gravels for comparison with those at the next stop.

7 Travel east about ½ mile to the first right turn, marked Dry Creek. Turn south on Dry Creek road (U.S. Forest Service road 280). The road reaches the mouth of Dry Creek in one mile. The gravel pit at the curve on the right side of the road is private, however, much can be observed from the road right-of-way.

8 **Stop 4: Dry Creek gravel pit**

The gravel deposit is at the mouth of Dry Creek. The landform trails southeast from the valley side in a direction opposite the flow of water that emptied glacial Lake Missoula (Figure 3). This pit is at an elevation of 817 meters, over 125 meters above the floor of the valley. The foresets (Figure 8) are several meters high and dip eastward. Clasts are poorly rounded and some cobbles are striated. Although most clasts are Precambrian Belt metasediments derived locally, some clasts of granite and diorite rocks exposed in and adjacent to the Purcell Trench are present in addition to a few clay balls. The sediments are very well cemented, and the pit walls must be ripped before the aggregate can be excavated.

These gravels probably represent glacial material deposited into glacial Lake Missoula. All the major drainages on the south side of the

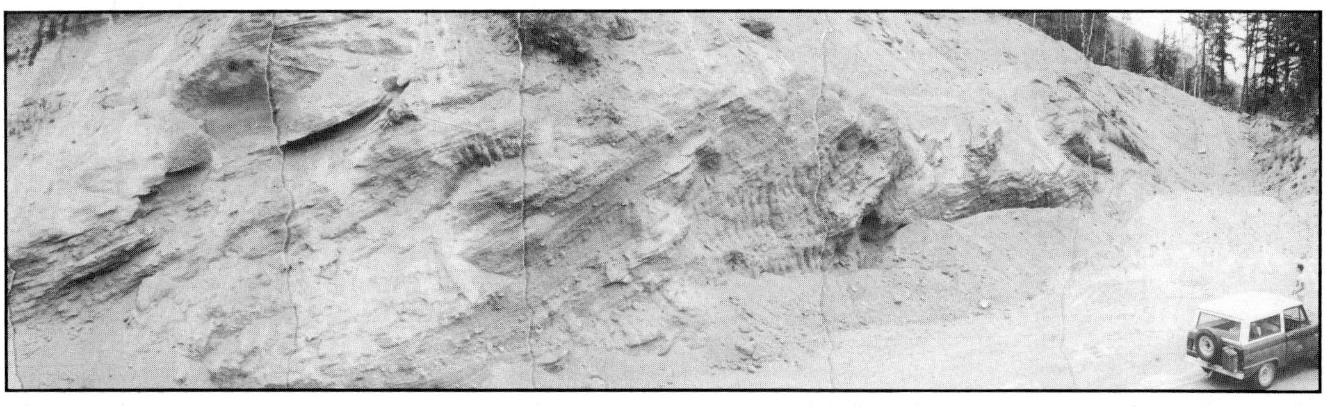

Figure 8. Panorama of Dry Creek gravel pit. Poorly rounded, unsorted gravels dip steeply to the east. These deposits probably were formed by outwash into glacial Lake Missoula.

Clark Fork valley between here and Trout Creek, over 30 kilometers to the southeast, have similar deposits. Known occurrences of these deposits are restricted to the south side of the Clark Fork valley opposite the source areas of ice on the north just where ice marginal channels would be expected. If these features represent ice marginal outwash deposited into Lake Missoula, then the ice extended much farther up the Clark Fork valley than previously recognized, nearly to Thompson Falls, and the late phases of glacial Lake Missoula drained quiescently enough to leave them preserved.

9 Return to road junction. Turn east and follow signs toward Heron.

11 Idaho-Montana state line. The route crosses flood gravels. Megaripples that are capped by lake silts are visible in many of the hay fields along the road. Bedrock hills in the valley have pendant bars trailing downriver.

At junction just before Heron, turn left and then right and cross the railroad tracks with care. Proceed across bridge over Cabinet Gorge Reservoir (Clark Fork River) to Montana State Highway (S.H.) 200 and turn left (west).

Milepost Description

1 Blue Creek bay. Deposits of Lake Missoula silts are exposed in the bay to the north.

63 Montana-Idaho state line. Turn south at the Cabinet Gorge Dam sign to the visitor's parking area. For a better view, if time permits, walk up the back (north) side of the prominent knob overlooking the dam site. **Caution! Take care in approaching the edge of the cliff.**

Stop 5: Cabinet Gorge Dam viewpoint.

Completed in 1952 by Washington Water Power Company (WWP). The 183-meter-long and 63-meter-high true arch dam is constructed on the Libby Formation of Precambrian Belt Supergroup. This dam, in coordination with the Albeni Falls Dam downstream, controls the water level of Pend Oreille Lake. A large new hatchery just downriver of the dam provides two million Kokanee salmon smolts to the lake annually.

The gravels capping the prominent terrace south of the river are mostly flood deposits, but logs from monitor wells drilled by WWP in the terrace show cycles of clay till and interbedded lake deposits indicating multiple episodes of ice damming (H.T. Stearns, consulting geologist, oral communication, 1986). Glacial erosion and till deposits indicative of an ice margin are common in this area, therefore many researchers have mapped the ice lobe terminus near here (see, for example, Alden, 1953; Weis and Richmond, 1965; Waitt, 1985).

Note the truncated spurs and remnant strandlines on the south side of the Clark Fork valley. The nearly straight front along the north side of the valley here is controlled by the Hope fault, which may have been active in the Quaternary. If so, it is possible that tectonic or isostatic activity may have had a role in the failure of the ice dam.

The bedrock bench on the north side of the valley has abundant till cover interpreted to be ice marginal deposits; possibly flood drainage was pushed to the south side of the valley by the ice from the north, at least in the waning stages of smaller late floods. Perhaps the bench represents the edge of the ice dam failure or margin of subglacial flow. Just to the north, ice flowed through cols as high as 1830 meters across the Cabinet Mountains between the Bull River and the Purcell Trench. Return to S.H. 200 and proceed west toward Clark Fork.

60 River Delta. Here ground ice was reported by highway construction crews (oral communication, Idaho District 1 office, 1987). Ground ice has been reported at several other highway construction sites between Clark Fork, Idaho, and Thompson Falls, Montana, and at Lakeview on Pend Oreille Lake (McLeod, 1987). Although sometimes called glacial ice, these occurrences probably represent recent seasonal buildup rather than Pleistocene remnants.

59 Large ripples in hayfield south of highway.

58 Just past milepost 58, a state-owned gravel pit is on the right. Imbricated Missoula flood gravels dip generally west and are overlain by silt.

56 Before entering the town of Clark Fork, large glacial grooves and striations can be seen in roadcuts in the Wallace and Striped Peak Formations along the curves in the road (shown in Figure 9; Bush and Breckenridge, in press).

Pass through the town of Clark Fork and return to Sandpoint. Then travel south toward Coeur d'Alene on U.S. 95 by same route taken coming north. For an alternate return to Spokane, turn right at Athol and proceed west on Idaho State Highway (S.H) 54 toward Spirit Lake.

From Athol to Spirit Lake, S.H. 54 crosses the large fan of flood gravels discharged from the end of Pend Oreille Lake. The road also crosses a large-scale ripple

Figure 9. Glacial grooves and striations on Precambrain Belt Rocks along S.H. 200 near Clark Fork.

field. This is one of the largest of the flood ripple fields and is most striking when viewed from the air. Flood water, discharged from what is now the end of Pend Oreille Lake Basin directed flood outbursts toward this area and formed the gravel dam that holds Spirit Lake. Bedding in the flood gravels dips toward the west, up the Spirit Lake basin.

At the junction of S.H. 54 with S.H. 41, turn south toward Rathdrum on S.H. 41. At Rathdrum continue west on S.H. 54 toward Trentwood, Washington. This route passes several large gravel bars that dam Twin and Newman Lakes. At Pines Road turn south and return to I-90, or proceed on Trent Avenue back to Spokane.

ACKNOWLEDGEMENTS

This field guide benefited from the helpful comments and suggestions of Eugene P. Kiver, Robert L. Logan, Fred K. Miller, and Loudon R. Stanford. I thank Roger C. Stewart for editing and Jennifer Pattison Hall for design and layout. Field studies were supported in part by the U.S. Geological Survey's surfical geology mapping project. I am grateful to Gerald M. Richmond, Kurt L. Othberg, and Robert R. Curry for comments in the field. Seismic information obtained from Pend Oreille Lake was supplied by the U.S. Navy David Taylor Research Facility at Bayview. I thank Ken and Edie Kinucan of the University of Idaho Clark Fork Field Station for their hospitality and interest in my research. Many residents of the Clark Fork area granted access to their land and provided useful information.

REFERENCES

Aitkenhead, N., 1960, Observations on the drainage of a glacier-dammed lake in Norway: Journal of Glaciology, v. 3, no. 27, p. 607-609.

Alden, W. C., 1953, Physiography and glacial geology of western Montana and adjacent areas: U.S. Geological Survey Professional Paper 231, 200 p.

Anderson, A. L., 1927, Some Miocene and Pleistocene drainage changes in northern Idaho: Idaho Bureau of Mines and Geology Pamphlet 18, 29 p.

Baker, V. R., 1973, Paleohydrology and sedimentology of Lake Missoula flooding in eastern Washington: Geological Society of America Special Paper 144, 79 p.

Bretz, J H., 1923, The channeled scablands of the Columbia Plateau: Journal of Geology, v. 31, p. 617-649.

Bush, J. H., Jr., and R. M. Breckenridge, editors, in press, Guidebook to the geologic history of northern Idaho and southeastern Washington: Field trips originating from Moscow-Pullman: Idaho Geological Survey Bulletin 29.

Chambers, R. L., 1971, Sedimentation in glacial Lake Missoula: University of Montana M. S. thesis, 100 p.

Clague, J. J., 1978, Mid-Wisconsin climates of the Pacific Northwest, in Current Research, Part B: Geological Survey of Canada Paper 78-1B, p. 95-100.

———, 1981, Late Quaternary geology and geochronology of British Columbia, Part 2: Summary and discussion of radiocarbon dated Quaternary history: Geological Survey of Canada Paper 80-35, 41 p.

Conley, Cort, 1982, Idaho for the curious, a guide: Backeddy Books, Cambridge, Idaho, 704 p.

Conners, J. A., 1976, Quaternary history of northern Idaho and adjacent areas: University of Idaho Ph.D. dissertation, 504 p.

Costa, J. E., 1988, Floods from dam failures, *in* V. R. Baker, R. C. Kochel, and P. C. Patton, editors, Flood Geomorphology: John Wiley and Sons, p. 439-463.

Craig, R. G., 1987, Dynamics of a Missoula flood, *in* L. Mayer, and D. Nash, editors, Catastrophic Flooding: Allen and Unwin, p. 305-332.

Craig, R. G., and J. P. Hanson, 1985, Erosion potential from Missoula floods in the Pasco Basin, Washington:

Battelle Pacific Northwest Laboratory Document PNL-5684, 266 p.

DeSmet, J. S., 1983, Geohydrology of the Southside aquifer, Sagle, Idaho: University of Idaho Master of Engineering Technical Paper, 25 p.

Dort, Wakefield, Jr., 1960, Glacial Lake Coeur d'Alene and berg-rafted boulders: Idaho Academy of Science Journal, v. 1, no. 2, p. 81-92.

Finckh, Peter, Kerry Kelts, and Andre Lambert, 1984, Seismic stratigraphy and bedrock forms in perialpine lakes: Geological Society America Bulletin, v. 95, no. 9, p. 1118-1128.

Flint, R.F., 1936, Stratified drift and deglaciation in eastern Washington: Geological Society of America Bulletin, v. 47, no. 12, p. 1849-1884.

Glenn, J. W., 1954, The stability of ice-dammed lakes and other water filled holes in glaciers: Journal of Glaciology, v. 2, no. 15, p. 316-318.

Liestøl, Olav, 1955, Glacier-dammed lakes in Norway: Norsk Geografisk Tidsskrift, v. 15, no. 3-4, p. 122-149.

McLeod, J. R., 1987, Mysterious Lake Pend Oreille and its "Monster": Fact and folklore: The Cryptozoology Club, North Idaho College, Woodcraft Publishing, Coeur d'Alene, 97 p.

McKiness, J. P., 1988, The Quaternary System of the eastern Spokane Valley and the lower, northern slopes of the Mica Peak Uplands, eastern Washington and northern Idaho: University of Idaho M. S. thesis, 98 p.

Molenaar, Dee, 1988, The Spokane Aquifer, Washington: Its geologic origin and water-bearing and water-quality characteristics: U.S. Geological Survey Water-Supply Paper 2265, 74p.

Newcomb, R. C., and others, 1953, Seismic cross sections across the Spokane River valley and the Hillyard Trough, Idaho and Washington: U.S. Geological Survey Open File Report 53-199, 16 p.

Pardee, J. T., 1910, The glacial Lake Missoula: Journal of Geology, v. 18, p. 376-386.

————, 1942, Unusual currents in glacial Lake Missoula: Geological Society of America Bulletin, v. 53, no. 11, p. 1569-1600.

Richmond, G. M., 1986, Tentative correlation of deposits of the Cordilleran ice sheet in the northern Rocky Mountains, *in* Vladimir Sibrava, D. Q. Bowen, and G. M. Richmond, editors, Quaternary Glaciations in the Northern Hemisphere: Quaternary Science Reviews, v. 5, p. 129-144.

Richmond, G. M., P. L. Weis, and Roald Fryxell, 1965, Part F: Glacial Lake Missoula, its catastrophic flood across the Columbia Plateau, and the loesses and soils of the Columbia Plateau, *in* C. B. Schultz and H. T. U. Smith, editors, Northern and Middle Rocky Mountains – Guidebook for field conference E, International Association for Quaternary Research, VII Congress: Nebraska Academy of Sciences, p. 72-73.

Savage, C. N., 1965, Geologic history of Pend Oreille Lake region in north Idaho: Idaho Bureau of Mines and Geology Pamphlet 134, 18 p.

Scott, W. E., K. L. Pierce, J. P. Bradbury, and R. M. Forester, 1982, Revised Quaternary stratigraphy and chronology in the American Falls area, southeastern Idaho, *in* Bill Bonnichsen and R. M. Breckenridge, editors, Cenozoic Geology of Idaho: Idaho Bureau of Mines and Geology Bulletin 26, p. 581-595.

Thorarinsson, Sigurdur, 1939, The ice-dammed lakes of Iceland with particular reference to their values as indicators of glacier oscillations: Geografiska Annaler, v. 21, Ht. 3-4, p. 216-242.

Waitt, R. B., 1985, Case for periodic colossal jökulhlaups from Pleistocene glacial Lake Missoula: Geological Society of America Bulletin, v. 96, no. 10. p. 1271-1286.

Waitt, R. B., Jr., and R. M. Thorson, 1983, The Cordilleran ice sheet in Washington, Idaho, and Montana, *in* H. E. Wright, Jr., editor, Late-Quaternary Environments of the United States, Volume 1, The Late Pleistocene: University of Minnesota Press, p. 53-70.

Webster, G. D., M. J. P. Kuhns, and G. L. Waggoner, 1982, Late Cenozoic gravels in Hells Canyon and the Lewiston basin, Washington and Idaho, *in* Bill Bonnichsen and R. M. Breckenridge, editors, Cenozoic Geology of Idaho: Idaho Bureau of Mines and Geology Bulletin 26, p. 669-683.

Weis, P. L., and G. M. Richmond, 1965, Maximum extent of late Pleistocene Cordilleran glaciation in northeastern Washington and northern Idaho: U.S. Geological Survey Professional Paper 545C, p. 128-132.

Subaqueous Basalt Eruptions Into Pliocene Lake Idaho, Snake River Plain, Idaho

Margaret D. Jenks[1]
Bill Bonnichsen[1]

INTRODUCTION

The central and western parts of the Snake River Plain provide classic exposures for studying the effects of water on the products of basaltic eruptions. Significant variations in water depth, from a large deep lake to an elevated water table, produced exposures with a wide range of characteristics. In addition, eruptions which ranged from small dike-fed flows to large, multi-phase volcanoes created a variety of volcanic products including flows that ran into the lake, pillowed basalt flows that formed lake-margin deltas and volcanic edifices, pyroclastic tuffaceous units deposited beneath varying water depths, and hydrovolcanic constructions including tuff cones, tuff rings, and maars. All of these varied deposits are clearly exposed in the canyon walls of the Snake River and its tributaries and in the surrounding desert terrain.

[1] *Idaho Geological Survey, University of Idaho, Moscow, Idaho 83843.*

The topographic lowland called the Snake River Plain formed as a result of two tectonic events. The western Snake River Plain is a graben that may have originated from extensional faulting similar to that in the Basin and Range province in Nevada and eastern Idaho. The central Snake River Plain is part of a trend of bimodal rhyolitic/basaltic volcanism and associated tectonism that extends from northern Nevada across the southern part of Idaho to the Yellowstone National Park area in northwestern Wyoming.

This two-day trip (Figure 1) will concentrate on two areas near the intersection of the western Snake River Plain graben with the northeast trend of bimodal volcanism. During the first day we will visit volcanoes and flows within and east of the Bruneau River canyon in order to review the history of the basaltic volcanism in that area in relation to Pliocene-Pleistocene Lake Idaho. We will spend the second day in the vicinity of Sinker Creek on the south side of the Snake River canyon, examining

Jenks, M.D., and Bill Bonnichsen, 1989, Subaqueous basalt eruptions into Pliocene Lake Idaho, Snake River Plain, Idaho, *in* V. E. Chamberlain, R. M. Breckenridge, and Bill Bonnichsen, editors, Guidebook to the Geology of Northern and Western Idaho and Surrounding Area: Idaho Geological Survey Bulletin 28, p. 17-34.

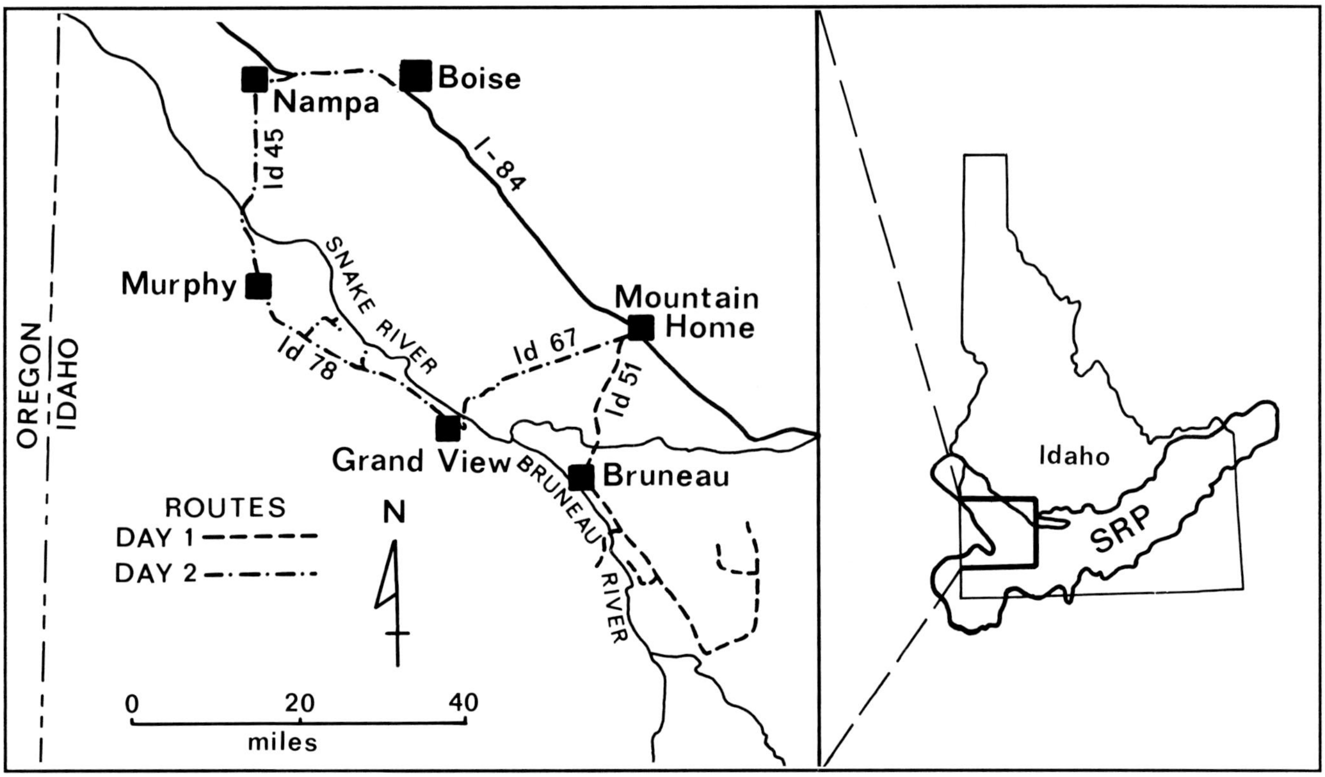

Figure 1. Field trip route in the central and western parts of the Snake River Plain, Idaho.

volcanoes and pyroclastic basaltic tuffs that also were erupted into or deposited within Lake Idaho. In this area we will observe the effects of topography as well as water on the basaltic eruptive products. Both days will offer participants excellent rock exposures for comparing a wide variety of depositional and eruptive structures associated with hydrovolcanic eruptions.

HISTORY OF THE LAKE IDAHO CONCEPT

The earliest geologic explorers of the Idaho Territory believed that during Tertiary time a large lake had occupied most of the western and central parts of the Snake River Plain. Members of both the Hayden expedition (Hayden, 1872) and the King survey of the 40th Parallel (King, 1878, 1903) collected fossil fish and mammal bones from the Snake River Plain and its environs. Cope (1870, 1883a, 1883b) examined the fish fossils and found them to be similar to those from other ancient large lakes in North America, such as Lake Bonneville and Lake Lahontan. Cope (1883a) described the ancient environment in which the fish had lived as fresh water lacustrine, and he hypothesized the existence of a large lake which he named Lake Idaho. More detailed analyses of the fossils and mapping of their accompanying sediments (Lindgren and Drake, 1904; Russell, 1902) confirmed the existence of this large Tertiary-age lake. Unconformities within the sedimentary record suggested two distinct basin-filling episodes, and the sediments were divided into the Payette and Idaho Formations (Lindgren and Drake, 1904).

Later workers found, however, that distinguishing in detail between the sediments of the Payette Formation and the Idaho Formation was difficult. Buwalda (1921) and Kirkham (1931) were puzzled by the variety of adjacent and seemingly contemporaneous sedimentary environments they found within the two formations in the vicinity of the Idaho-Oregon border. They decided to discard the "large, deep, long-standing lake" theory in favor of numerous, temporary shallow lakes of varying sizes. Stearns (1938) working at the same time in the central Snake River Plain retained the "large, deep lake" theory in his mapping of the Hagerman Lake Beds. Later geologists (Malde and Powers, 1962; Malde and others, 1963; Malde, 1972), mapping primarily the sedimentary rocks in the central plain, agreed with Buwulda and Kirkham. Malde's paper (1972) on the depositional environments in the Glenns Ferry Formation states the problem succinctly:

> Evidently, the river flowed in a wide valley marked by temporary lakes and by broad

> stretches that were seasonally flooded. As the river shifted its course, the sedimentary environments changed correspondingly. Even so, the persistence of rather uniform environments in certain areas is shown by surprisingly thick sequences of fairly uniform deposits.
>
> This implied persistence of sedimentary environments presents difficult problems about the paleogeography that are still unresolved. The great thickness of the lacustrine facies in a wide area around Glenns Ferry is particularly perplexing, especially because a considerable part of it evidently accumulated while flood-plain sediments were being deposited nearby.

Recently, the concept of Lake Idaho as a large permanent lake has been revived by geologists working in the central and western plain. Paleontologists (Miller and Smith, 1967; Smith, 1975; Smith and others, 1982) studying fossil fish morphology and evolution concluded that the fish lived in large and deep lake environments. Swirydczuk and others (1982) used fission track dating of volcanic ash layers within the lacustrine and associated fluvial and deltaic sediments to correlate stratigraphic sections. Kimmel (1982) and Middleton and others (1985) analyzed depositional environments from eastern Oregon to the central plain near Twin Falls, Idaho, in order to gain a broader view of the tectonic setting and geologic history of the western Snake River Plain lake. These excellent, detailed studies have contributed much to our understanding of the history of the sedimentary rocks in the western Snake River Plain, but they did not include a study of the basaltic volcanic rocks.

BASALTIC VOLCANISM AND THE STUDY OF LAKE IDAHO

Previous researchers in the Snake River Plain found large volumes of basaltic volcanic materials within the sediments and recognized the existence of phreatic as well as subaerial volcanoes. Geologists studying the sediments of the western and central Snake River Plain developed a stratigraphic framework (Malde and Powers, 1962; Malde and others, 1963) that dealt with the volcanic rocks in two ways. The basalt flows either were lumped into large formations that contained many flows from different volcanoes (e.g., the basalt of the Snake River Group and the Banbury Basalt) or were included as members or interbeds within sedimentary lithostratigraphic units (e.g., the Chalk Hills basalt member of the Chalk Hills Formation). Mappers using the lithostratigraphic approach included the flows from a single volcano in more than one stratigraphic formation. We suggest that this lithostratigraphic approach, which is valid for the sedimentary units, has limited value in the Snake River Plain because it does not recognize the chronostratigraphic importance of the basaltic lithologies. In addition, the introduction of large amounts of basaltic materials into a sedimentary system may drastically change the depositional sedimentary environments. These changes, which were not recognized by the earlier workers who created the lithostratigraphic units, may explain some of the apparent complications in correlating sedimentary sequences in the western Snake River Plain.

Our recent stratigraphic research has focused on the basalt units and their sources in the central and western Snake River Plain (Jenks and Bonnichsen, 1987; Jenks and Bonnichsen, in press(a); Jenks and Bonnichsen, in press(b); Bonnichsen and Jenks, in press; Jenks, Bonnichsen, and Godchaux, in prep.). By mapping chronostratigraphic volcanic units, we are attempting to create a new volcanologic framework for interpreting the geologic history of the area. From this research we have concluded that a series of large permanent lakes, similar to the Lake Idaho envisioned by Cope and other earlier geologists, did in fact occupy the same depositional basin in the western and central Snake River Plain during the late Tertiary and early Quaternary. During this time there were recessions of the lakes and periods of erosion, but we have chosen to use the term "Lake Idaho" to stand for all of these episodes, much as "Lake Bonneville" is used for the series of lakes that occupied the Bonneville basin in Utah.

We believe that the basaltic volcanism was affected in major ways by its interactions with the Lake Idaho lakes and that, conversely, the sedimentary depositional environments were affected by the eruption of subaqueous volcanoes and the influx of large amounts of basaltic material into the lakes. The interaction of the water with the basalt volcanoes and flows created several different kinds of volcanic products depending upon the timing of the basalt/water interaction. Water-affected and pillowed flows formed when subaerial lavas reached the lakeshores. Hydroclastic volcanoes erupted directly into the lakes and produced bedded basaltic tuffs and near-vent packages of explosive volcanic debris. The following discussion describes our present understanding of these volcanic products.

Water-Affected Basalt Flows

We believe that many of the basalt flows in the Snake River Plain were changed, but not pillowed, when they entered the Lake Idaho lakes. We arrived at this conclusion during our unit-by-unit mapping of the basalt flows in the Bruneau River canyon area (Jenks and Bonnichsen, in press(a); Bonnichsen and Jenks, in press). Here we demonstrated that single-source packages of flows were separated by intervening sedimentary layers. In numerous places we observed that these sediments were "baked" to a red color and hardened by the heat of the overriding flows. Earlier workers (Malde, and others, 1963) had observed what appeared to be more weathered basalt flows and had mapped them as parts of older units. Using the red sediment layers as marker beds, we were able to show that basalt units, which were previously thought to be older, are in fact the lateral stratigraphic equivalents of subaerial and unweathered flows. The weathered-or older-appearing exposures and the subaerial flows are actually parts of the same unit, with the weathered appearance being due to an interaction with water. In this paper we have used the term "water-affected basalt" to refer to the parts of the units that show these weathering patterns.

We believe that the water-affected basalt flows were not pillowed because they flowed into the lakes over relatively low-relief shorelines. This allowed them to keep their internal coherence. High eruption rates may be another cause of this apparent coherence. Some parts of the flows appear to have been chilled by the water and thus have remained relatively unaltered. Other parts, however, reacted with the water and were markedly altered, thus becoming susceptible to subsequent internal alteration and disintegration. When fully-developed, the water-affected parts of the flows weather to a brown, granular, disintegrated rock—basaltic grus—which encloses zones of chilled, unaltered basalt. The unaltered basalt within the brown basaltic grus suggests that perhaps parts of the flows were protected from the effect of the water by factors such as flow thickness, ponding, distance from cooling fractures, or slow cooling rates.

In Jenks and Bonnichsen (1987) we documented several areas around the central and western parts of the Snake River Plain where up to twelve successive units of water-affected basalt flows are exposed. These include the lower Bruneau River area, Melon Valley, the King Hill area, and eastern Oregon near the Idaho-Oregon line. The regional exposure of water-affected basalt flows, their total thickness, their time-transgressive quality, and their restriction to within the Lake Idaho basin are further evidence that the principal cause of their alteration was interaction with the waters of the Lake Idaho lakes.

Pillowed Basalt Flows

The second effect of water on the basalt flows was pillowing. Pillowed basalt flows are not as numerous in the western Snake River Plain as water-affected flows. We believe that pillows formed where basalt flows poured down a relatively steep slope, and pillow deltas appear to have formed along steeper shorelines. In two places (Pence Butte in the lower Bruneau River area and Hill 3337 (elevation) in the Sinker Creek area) volcanoes built platforms or piles of pillows as part of their volcanic construction.

Pillowed flows are numerous in the Sinker Creek area where they form the lower zones of many basalt units (Jenks and Bonnichsen, in press (b); Jenks, Bonnichsen, and Godchaux, in prep.). The Nahas Ranch basalt is of particular interest, having a thick lower pillowed zone behind an apron of tuffaceous detritus that was apparently shed off the fronts of the advancing flows. Other areas in the western and central Snake River Plain of pillowed basalt flows include the Lily Grade near Castleford (Bonnichsen, and others, 1988), the Clear Lakes grade near Buhl, and the lower Bruneau River canyon, Stop 2 on the first day of this trip.

Massive and Bedded Basaltic Tuffs

Massive and layered tuffs of fresh and palagonitized pyroclastic materials are the third type of volcanic product formed by the interaction of basaltic volcanism with the waters of the Lake Idaho lakes. Reaching thicknesses of 600 feet near the volcanoes, exposures of these units appear to be limited to within a few miles of their sources. Thick sequences of tuff also are present where contemporaneous materials from the eruption of adjacent volcanoes are interlayered. None of these deposits have been studied in detail. However, we have identified many interesting depositional and eruptive structures in these hydrovolcanic tuffs, including flame structures, dewatering channels, rounded and porcelaneous clasts of lake sediments, graded bedding, disrupted bedding, rip-up clasts, clastic dikes, accretionary lapilli, and cross-bedded layers deposited by base surges or small turbidity-type slumps. Additional puzzling features of the tuffs include alternating palagonitized and unaltered cinder layers, zones of welded basaltic tuff, and both sedimentary and volcanic lithic clasts.

Although bedded pyroclastic materials can be found throughout the central and western parts of the plain, they appear to be most abundant and have the thickest exposures in the area surrounding the Snake River near Sinker Creek. During the second day of the trip participants will observe some of these features in the Red Trails tuff at Stop 10.

Hydroclastic Constructions

The fourth type of volcanic product caused by the interactions of basaltic lavas with the Lake Idaho lakes are the volcanoes that erupted underwater or had major explosive interactions with water. Hydrovolcanic constructions and hydroclastic volcanoes are located throughout the central and western parts of the plain within the Lake Idaho basin, including the two areas visited by this trip. In the lower Bruneau River area these include Pot Hole Butte, Pence Butte, Sailor Cap Butte, and 71 Gulch volcano. In the Sinker Creek area Jackass Butte, Castle Butte, Fossil Butte, Hill 3337 (elevation), Sinker Creek Butte, Montini volcano, Sinker Butte, Con Shea Basin volcano, and Guffey Butte as well as several unnamed volcanoes all contain substantial quantities of hydroclastically erupted materials.

We will visit four hydroclastic volcanoes on this trip to observe a range of volcanic features. The near-vent deposits of the underwater eruptions vary from thick accumulations of pillows to massive, unsorted, palagonitized and fresh, cindery tuff to thick stacks of water-affected basalt flows. The range of eruptive and depositional structures appears to be dependent on a number of factors including eruptive rates, lava viscosity, and the amount of water interaction and contained volcanic gases. Several of the volcanoes built edifices that eventually reached the subaerial/subaqueous interface, thus changing their eruption mode to subaerial flowage. Others developed lava ponds that were protected from the water by crater walls. One volcano, Sinker Creek Butte (Stop 8), appears to have had its vent inundated by water during eruption, because its oldest products are subaerial flows that are pillowed or water-affected on its flank and its youngest products are hydroclastic tuffs.

PRESENT EVIDENCE FOR LAKE IDAHO

We believe that evidence from the basalt volcanism documents the presence of Lake Idaho as a series of large permanent lakes in the central and western parts of the Snake River Plain in the late Tertiary. The following discussion outlines our present understanding of Lake Idaho's boundaries and geologic history.

The geomorphological and sedimentary shoreline features of the Lake Idaho lakes have been altered by later erosional and depositional processes and so are not as obvious as those of younger lakes like Lake Bonneville. However, the Lake Idaho shorelines can be defined using the geographic limits of the basalt-water interaction. In the lower Bruneau Canyon as well as in other areas throughout the central and western plain, features in the basalt flows suggest a Lake Idaho highest stand at approximately the present 3800-foot elevation contour. Although it is probable that the shorelines of the earlier Lake Idaho lakes changed as the western Snake River Plain graben formed, the basalt stratigraphy in the lower Bruneau River area suggests that the 3800-foot elevation high stand in that region has not been materially altered by tectonics. Figure 2 shows the approximate outline of

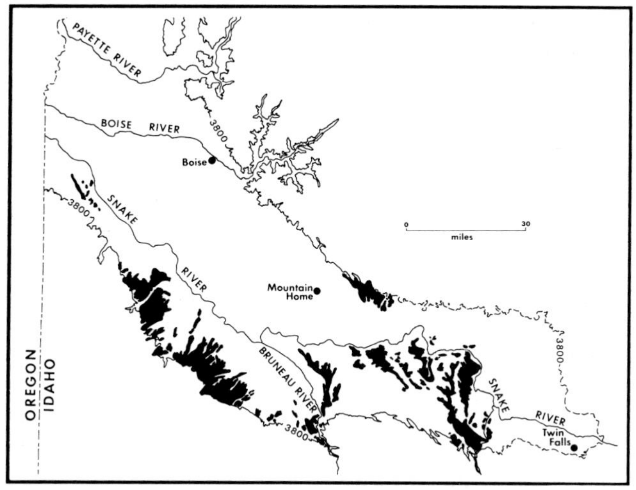

Figure 2. Approximate outline (using the 3800-foot contour) of the high-stand shoreline of Lake Idaho in the central and western parts of the Snake River Plain. The dashed line shows areas where the shoreline is covered by younger basalt flows. The dark areas are exposures along the southern edge of the Lake Idaho basin of gravels which cap the fine-grained lacustrine sediments. The gravels begin at approximately the 3800-foot contour and extend downslope to approximately the 2900-foot contour. They were probably deposited by braided streams or alluvial fans, or as deltaic or beach deposits related to the ancestral drainages of the rivers and streams that flowed into the lake. Compilation based on Ekren and others (1981) and Malde and others (1963).

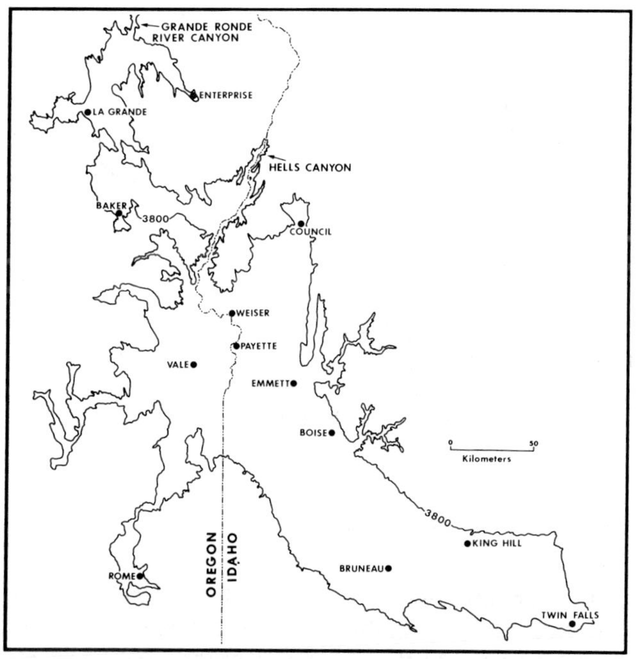

Figure 3. Approximate outline of the Lake Idaho basin and other Tertiary lacustrine basins in eastern Oregon and western Idaho. The outline is based on our postulated 3800-foot, high-stand elevation for the Snake River Plain portion of Lake Idaho, and it follows a major regional topographic break.

the Lake Idaho lakes at their highest stand. If that elevation is extrapolated regionally in eastern Oregon and southwestern Idaho (Figure 3), it follows an approximate regional topographic break that outlines a number of Tertiary lacustrine basins.

An adequate framework of radiometric age dates for the volcanic units in the central and western Snake River Plain is not yet available. Published dates (Amini and others, 1984; Ekren and others, 1984; Hart and Aronson, 1983; Armstrong and others, 1975; Armstrong and others, 1980) suggest that the lake basin was filled in the Miocene and that the final recession of the lakes may have been in the early Pleistocene. Unconformities within the basaltic stratigraphic sequence, paleotopographic irregularities that were buried by basalt flows, and angular unconformities within the Idaho Group sediments show a series of recessions during which the basin sedimentation alternated between lacustrine, fluvial, and fluvial-deltaic depositional environments. A detailed geologic history of the Lake Idaho lakes awaits a more complete framework of radiometric age dates and geologic mapping that integrates the volcanic and sedimentary stratigraphies.

Moderate to extremely low-relief shorelines for the Lake Idaho lakes are suggested by the presence of water-affected rather than pillowed lava flows, of algal limestone reefs (Jones, 1978), and of extensive oolitic carbonates (Swirydczuk and others, 1979, 1980; Gallegos and others, 1987). The chemically precipitated sediments also suggest that parts of the lake basins had low sedimentation rates. Conversely, mono-layers of cobble-sized gravel (Jones, 1978) suggest that floods may have periodically moved shore gravels into the lake basin or upland gravels onto the lakeshore.

It appears that most of the canyon-cutting of the Bruneau and Jarbidge Rivers and other streams that emptied into the Lake Idaho basin occurred during and after the final lake recession. This final lake recession appears to be represented by gravel sequences that cap the lacustrine deposits in many areas (Figure 2). The gravels may be the result of stream deltas or braided streams spreading across the basin as the final lake receded. The distribution and configuration of these gravels suggest that they are related to the present drainages, either as alluvial fans built from the front of the Owyhee Range or as deltas and shoreline features related to the ancestral drainages of the Bruneau River, Jarbidge River, Salmon Falls Creek, Clover Creek, and Sheep Creek. Our reconnaissance of the gravel deposits in the area around Bruneau River canyon found that they are nearly devoid of basaltic clasts, despite the fact that the streams depositing these gravels flowed north from Nevada over more than 50 miles underlain almost exclusively by basalt flows. We account for this by suggesting that the gravels were deposited by braided streams and that little downcutting or incision was occurring in this drainage before the final Lake Idaho lake receded and the base level changed.

ROAD LOG

Day 1: Geology of the Lower Bruneau River Area

The field trip route for Day 1 follows the east side of the lower Bruneau River canyon (Figure 4). The roads are mainly dirt and gravel, and they may not be passable from early winter to early spring. A four-wheel drive vehicle is not essential, but a vehicle with adequate clearance is recommended. Only Stops 1 and 5 are accessible to commercial buses. Individuals attempting the trip on their own may wish to obtain the appropriate 7 ½-minute quadrangle maps, because some of the roads do not appear on either the 15-minute quadrangle maps or the Twin Falls 1 x 2 degree sheet.

The trip route passes close to the western and southern sides of the Saylor Creek Air Force Bombing Range (Figure 4) used by the Mountain Home Air Force Base. Please heed the warning signs posted along the Clover-Three Creek Road which crosses the approach flight path to the range. An example is: "WARNING. This road crosses a U.S. Air Force Bombing Range for the next 12

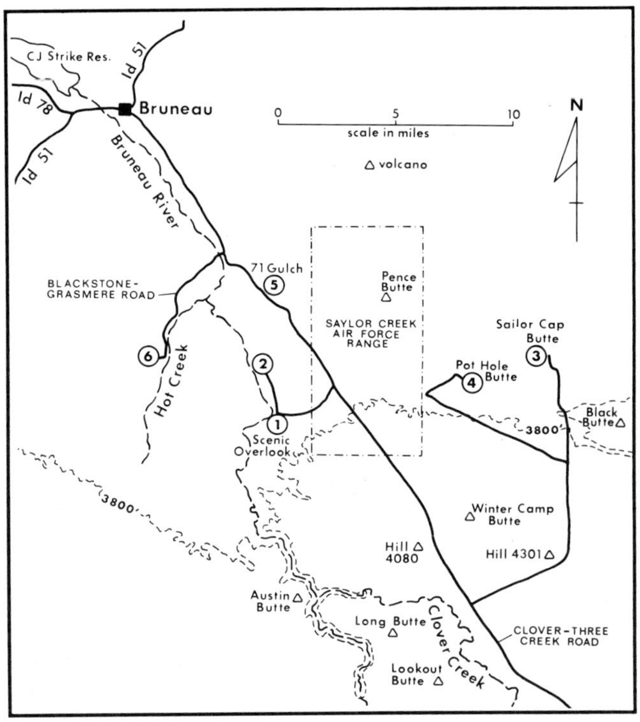

Figure 4. Trip route and stops for Day 1.

miles, dangerous objects may drop from aircraft." The target area of the range is fenced and off limits to civilians.

The mileage for the trip begins in Mountain Home, Idaho. Milesages are to the closest 0.05 mile. Interval mileages are shown in parentheses at the end of descriptions.

Mileage	**Description**
0.0	Intersection of business loop Interstate Highway 84 (I-84) and Idaho Highways 67/51 in Mountain Home. Turn west on Idaho 67/51 toward Mountain Home Air Force Base. *(1.25)*
1.25	Turn left (south) on Idaho 51 toward Bruneau, Idaho. *(19.35)*
20.6	In Bruneau turn left (south) on Hot Springs Road (Clover-Three Creek Road). Follow the road sign pointing to "Bruneau Canyon, 18 mi." *(7.7)*
28.3	Pass the junction with the Blackstone-Grasmere Road (pavement ends). Continue driving south. *(5.8)*
34.1	Junction near Bench Mark 3539 with the road to the Saylor Creek Bombing Range target area. Take the right fork. *(2.5)*
36.6	Turn right (west) on the Bruneau Canyon Scenic Overlook road. *(3.2)*
39.8	**Stop 1: Bruneau Canyon Scenic Overlook.**

Points of interest: characteristics of subaerial basalt units.

The Bruneau River Canyon at this point is approximately 700 feet deep. The lava flows exposed here are typical of subaerial flows in the central and western Snake River Plain. The twelve basalt units are separated by tan, brown, or red sediment layers (Figure 5). Most of the flows result from sheet flowage of pahoehoe lava which produced a flow-on-flow configuration of thin (less than 10 meters) flows. Lava tubes and squeeze-ups are rare, and the marginal zones of the flows are characterized by wide, flat lobes. Several of these lobate forms are visible in the west wall of the canyon.

The uppermost unit (on which you are standing) is the Winter Camp Butte basalt, which contains abundant plagioclase phenocrysts up to 1 centimeter long. This unit was erupted from Winter Camp Butte volcano, 9 miles to the southeast (Figure 4), and appears to have flowed to the northeast, north, and northwest.

Figure 5. View at Stop 1 of the Bruneau River canyon looking north (downstream). Twelve subaerial basalt units are exposed in the walls of the canyon at this point. The section is capped by the Winter Camp Butte basalt. Its terminus which contains a pillow delta, is visible on the west side of the canyon, as a rim above a slope in the middle of the photograph.

Throughout its exposure the Winter Camp Butte basalt is above a fairly thick (up to 10 meters) layer of clay- and silt-sized sediments. The basalt units beneath the Winter Camp Butte basalt were also erupted from shield volcanoes located to the south and east, but their flows appear to have been constrained to flow toward the northwest, following the downfaulted margin of the Sheep Creek rhyolite lava flow to the south.

Turn left (north) on the dirt road that intersects the east side of the Scenic Overlook road turnaround. This road and its continuation to the south are on a gravel surface. This gravel is exposed at elevations of 3600-3800 feet and covers a broad area on both sides of the canyon (Jenks and Bonnichsen, in press(a)). The dominant lithologies in the gravel are rhyolite pebbles and cobbles, and pebbles, cobbles, and boulders of a gold- to tan-stained quartzite; basalt clasts are rare. We tentatively interpret these gravel deposits as resulting from prograding braided streams or beach and delta environments produced as the ancestral Bruneau River flowed into a Lake Idaho lake. *(0.9)*

40.7	Take left (west) fork in the road in the northeast part of section 30, T. 8 S., R. 6 E. (Crowbar Gulch quadrangle). *(1.15)*
41.85	**Stop 2: Winter Camp Butte basalt pillow delta.**

Points of interest: basalt pillows, prograding pillow delta, water-affected basalt flows.

Park the vehicles at the top of the slope and walk west toward the canyon rim. Across the

Figure 6. Foreset beds in the pillow delta that is the terminus of the Winter Camp Butte basalt (Stop 2). This view is of the west side of the Bruneau River canyon.

canyon is a pillow delta that formed as the upper flows of the Winter Camp Butte basalt flowed into a Lake Idaho lake (Figure 6). Beneath the foreset beds of pillows is a slope containing several earlier water-affected basalt flows and underlying sediments. The ends of these older Winter Camp Butte basalt flows created a fairly steep slope, which may have caused the formation of the pillow delta.

Walk north to the edge of the Winter Camp Butte basalt on the east side of the canyon. Here the pillows are up to 6 feet long and 2 feet wide and have multiple glassy rinds. From this viewpoint the water-affected basalt flows of the Hot Creek basalt, which is beneath the Winter Camp Butte basalt, are visible. The Hot Creek basalt flows make the transition from subaerial to water-affected at a point just downstream from the Winter Camp Butte basalt terminus. The water-affected flows are exposed on the west side of the Bruneau River canyon as a dark layer above a red sediment. The Hot Creek basalt water-affected flows will be viewed at Stop 6. *(2.15)*

44.0 Return to the Scenic Overlook road. Turn left (east). *(3.2)*

47.2 Return to the Clover-Three Creek Road. Turn right (south). This road passes Winter Camp Butte volcano on the west (Figure 4). Winter Camp Butte is one of the largest shield/volcanoes in the lower Bruneau River area with a 281-foot height, a 3.75-mile diameter, and a 2.9-percent average side slope (Jenks, 1984). The road then traverses a small graben. Most of the basalt flows in this part of the central Snake River Plain are cut by normal faults. The small graben that the road crosses is 100 feet deep and extends for several miles to the southeast and northwest. Barely visible to the right in the bottom of the graben is a small shield volcano (Hill 4080, elevation). *(11.4)*

58.6 Turn left (east) on the road at elevation point 4061 in section 10 of the Winter Camp 7 ½-minute quadrangle. A sign at the intersection points in the direction you are to turn and reads: "Crows Nest 15, Hammett 38." The road skirts the south and east sides of Hill 4301 (elevation), a fairly large shield volcano located directly southeast of Winter Camp Butte volcano on a line parallel with the trend of the normal faulting. Hill 4301 (elevation) is surrounded by the flows from the Winter Camp Butte volcano. The volcano is 221 feet high and 2.2 miles across; it has an average side slope of 3.9 percent and a 3600-foot-diameter crater at its top (Jenks, 1984). *(8.8)*

67.4 Note a major junction at elevation point 3849 feet in section 8 of the Black Butte West 7 ½-minute quadrangle. Continue north. *(2.3)*

69.7 Junction with a road leading to a microwave installation on the top of Black Butte to the east (Figure 4). Black Butte is another large shield volcano, whose southwest flank has been cut by a normal fault with 200 feet of displacement. The volcano is elongate northwest-southeast, 327 feet tall, and 1.8 miles in diameter. It has a 2400-foot-diameter crater. *(2.8)*

72.5 **Stop 3: Sailor Cap Butte volcano.**

Points of interest: depositional and eruptive structures in eroded basaltic tuff.

Park on the north side of the butte in the flat area to the left of the road and walk northwest to the ridge of basaltic tuff.

Sailor Cap Butte volcano which is between 3500 and 3600 feet in elevation, is a hydroclastic volcano. It was probably erupted into or near the edge of a Lake Idaho lake. The central area of the volcano, Sailor Cap Butte proper, is a massive basalt and probably the remnant of the lava pond that filled the volcano crater. Surrounding and dipping away from this central area on three sides are the remnants of the ejecta blanket created by the eruption. The eroded remnant of the ejecta blanket (Figure 7) includes the following features: (1) layers of palagonitized, cinder-cored, accretionary lapilli; (2) layers and pods of fresh cinders with individual pieces up to 2 inches in diameter – some have no matrix, but others have a palagonitized, basaltic tuff matrix; (3) large chunks and blocks of non-

Figure 7. View of the palagonitized beds, which dip away from the central massive basalt of the Sailor Cap Butte volcano (Stop 3). These beds contain chilled and cindery blocks and bombs and lapilli within a fresh or palagonitized cindery matrix.

vesicular, cold-emplaced basalt; (4) large-scale, surge-type cross-bedding; and (5) sparse pumice and lithic clasts. In the area between the central butte and the eroded ring, brownish sand is present. This sand may be the original lake sediments or may be eolian in origin; the sandy sediments underlying the tuff are exposed in one area. To the west of the volcano are several smaller eruptive points along an east-west trending dike.

Follow the eroded remnant of the tuff ring to its northeast end and return to the vehicles. Retrace the trip route to the junction at elevation point 3849 feet at the same location as mileage point 67.4. *(5.1)*

77.6 Turn right (west) on the road at elevation point 3849 feet. Note: if weather conditions are inclement or threatening, Stop 4 should be omitted, because the roads are not well-traveled and may be impassable when wet. *(6.9)*

84.5 Turn sharp right (north) on the road in section 29 of the Pot Hole Butte 7 1/2-minute quadrangle. (If you drive up a slope and reach the gate in the east fence of the bombing range, you have gone too far. The fence is approximately 0.2 mile beyond the correct turn.) *(1.5)*

86.0 Turn right (north) at the junction in section 29 just before a small, ephemeral pan lake. Follow the track to the center of Pot Hole Butte and park. *(0.6)*

86.6 **Stop 4: Pot Hole Butte volcano.**

Points of interest: the central area of a hydroclastic volcano.

Pot Hole Butte is a hydroclastic volcano like Sailor Cap Butte volcano, but it does not have a central lava pond and only the central area of the volcano remains. The top of this volcano is located at 3743 feet in elevation, and its hydroclastic nature suggests that the water level in the Lake Idaho lake was near this height during its eruption. Pot Hole Butte is elongate northwest-southeast and approximately 0.5 mile long and 0.3 mile wide. The rim visible to the southeast may be flows of Winter Camp Butte basalt which stopped against the now-eroded south flank of the Pot Hole Butte volcano.

The central area of the volcano appears to consist of three distinct zones, all of which dip toward the middle of the exposure. A lower zone contains one or more flows of glassy basalt which weather to a slope. Above this is a zone similar to the layers in the eroded ejecta ring of the Sailor Cap Butte volcano. This middle zone contains accretionary lapilli and fresh and palagonitized cinders as well as blocks and bombs of nonvesicular cold-emplaced basalt in a palagonitized matrix. The upper zone may have been erupted subaerially and consists of small welded spatter and lava flows and numerous fusiform bombs and clots. The basaltic rocks of this upper zone contain numerous large plagioclase laths.

Return to the vehicles and retrace the route to the main road. *(9.0)*

95.6 Turn right (south) at elevation point 3849 feet, the same location as mileages 67.4 and 77.6. *(8.8)*

104.4 Turn right (north) on the Clover-Three Creek Road. At this intersection two more shield volcanoes, Long Butte volcano (on the right) and Lookout Butte volcano, are visible to the west. Long Butte volcano is almost buried by the flows from Lookout Butte volcano. However, 200 to 400 feet of its flows are visible in the Bruneau River canyon. The flows from Lookout Butte volcano, like those of Winter Camp Butte volcano, overlie a fairly thick section of sediments. Lookout Butte volcano is elongate north-south, 1.7 miles across, and 194 feet high; its flank slope averages 4.5 percent (Jenks, 1984).

After passing Winter Camp Butte, a large hill is visible to the north at approximately 1 o'clock. This is Pence Butte, another volcano, which is the principal target for the pilots using the Saylor Creek Air Force Bombing Range. It was used as a target for live aerial bombing using three-foot bombs during World War II, so its original top may have been blasted away. The butte itself is only 100 feet high, but it appears that it has been partially buried by lake sediments and younger flows so that the actual volcanic edifice is quite large. Its exact boundaries have not been mapped, but a thick (300-foot) section of pillowed basalt from Pence Butte volcano exposed in the West Fork Browns Creek suggests that it built a large edifice within a Lake Idaho lake. Pence Butte basalt has a distinctive lithology of cumulophyric rosettes of large (up to 1.5 centimeters) plagioclase laths and (up to 0.25 centimeter) olivine phenocrysts. We have not found this lithology in the basalt flows exposed in the lower Bruneau River canyon. *(16.05)*

120.45 **Stop 5: View of 71 Gulch volcano.**

Points of interest: hydroclastic volcano, dike system.

Park in the small gravel pit on the right (east) side of the road. Walk north to the top of a small hill (elevation point 3009 feet) to view 71 Gulch volcano to the east (Figure 8).

71 Gulch volcano, which appears to have erupted into a Lake Idaho lake, is a series of small eruptive points on a northwest-southeast trending feeder dike. The top of the main eruptive point is predominantly cinders, but other eruptive products include pillow lavas and pahoehoe toes, bedded basaltic ash and bombs containing inclusions of silicic pumice, and a northwest-trending brecciated dike containing large pieces of porcelaneous lake sediments. *(2.4)*

122.85 Turn left (west) on the Blackstone-Grasmere Road, just before the pavement begins. Cross the wooden bridge over the Bruneau River. Note: this bridge has restricted load limits and cannot be used by commercial buses. *(0.75)*

123.6 Turn left (south) and ascend a grade. *(2.75)*

126.35 Rim of Hot Spring Limestone. This reefal carbonate of algal origin covers 80 square kilometers and is one of the principal marker horizons in the lower Bruneau River area (Jones, 1978). It consists of successive layers of cylindrical algal colonies interspersed with areas filled with gastropod shells and sediment. Our mapping shows that it is stratigraphically above the Hot Creek basalt and beneath the Winter Camp Butte basalt (Jenks and Bonnichsen, in press(a)). *(1.15)*

127.5 **Stop 6: Hot Creek basalt.**

Points of interest: Hot Spring Limestone, water-affected basalt flows, "baked" red sediment layer.

Turn right (west) on the jeep trail opposite the bottom of the draw containing a desiccated vehicular carcass (dead junkalope). Park and walk up the jeep trail and then climb to the rim on your right (north). Walk down the east-facing slope to view the exposed units.

In this area a section of well-developed, water-affected Hot Creek basalt is exposed beneath sediments capped by the Hot Spring Limestone. The Hot Creek basalt, in turn, is above a well-exposed layer of red "baked" sediment. The water-affected basalt flows are almost

Figure 8. Panorama looking east at the west side of the 71 Gulch volcano (stop 5). The dike, which probably fed the cindery eruption that caps the hill, is visible within the lake sediments to the right and as scattered solid outcrops on the lower part of the hill to the left.

entirely weathered to a dark brown to reddish brown basaltic grus containing clasts of chilled basalt as well as a few larger areas of unaffected basalt. Directly above the top of the basalt is 50 to 100 feet of sediment containing cobble- and pebble-sized gravel overlain by siltstone with some clay and sand layers. Armstrong and others (1975) report whole rock K-Ar age dates of 8.4 ± 0.7 and 8.8 ± 0.5 Ma for the basalt of the Chalk Hills Formation, the equivalent of our Hot Creek basalt unit. *(32.2)*

159.7 Return to Mountain Home via Bruneau and Idaho 51.

Day 2: Geology of the Sinker Creek Area

On this day of the field trip we will visit several places in the Sinker Creek area to see excellent exposures of bedded pyroclastic tuffs and hydroclastic volcanoes. Both the age dates on the flows of the volcanic units (Amini and others, 1984) and the unconformities in the sediments exposed in the Sinker Creek area suggest that a period of lake regression and erosion occurred before several of the basaltic units were erupted. We will visit an older volcano, Castle Butte, as well as the younger Montini volcano and Red Trails tuff.

The trip route for Day 2 (Figure 9) is easily accessible to cars and vans, but Stops 9 and 10 are not suitable for commercial buses. The trip crosses two private ranches, the Nettleton Ranch on Castle Creek and the Nahas Ranch on Sinker Creek. Permission to enter should be obtained in advance by calling the owner of the Nettleton Ranch and the manager of the Nahas Ranch. In addition, Stops 8, 9, and 10 are within the boundaries of the Snake River Birds of Prey Area. Participants should avoid disturbing the raptors and their nesting areas. Again, the trip route begins at the intersection of business loop I-84 and Idaho 67/51 in Mountain Home.

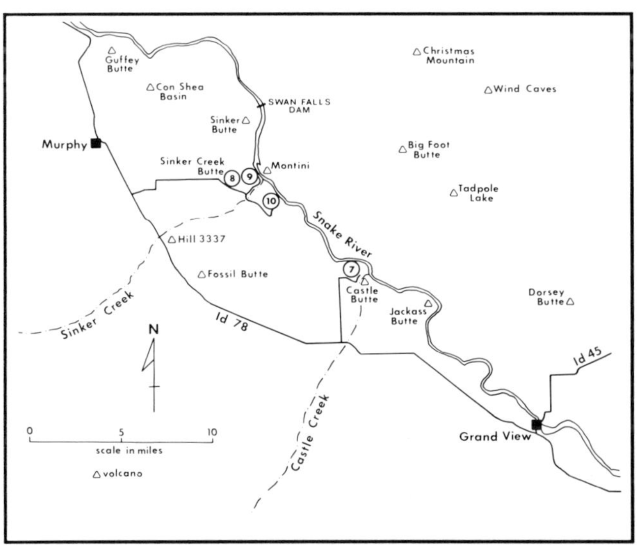

Figure 9. Trip route and stops for Day 2.

Mileage Description

0.0 Turn right (west) on Idaho 67/51. *(1.25)*

1.25 Junction of Idaho 51 and Idaho 67. Continue west on Idaho 67. *(7.65)*

8.9 Junction of Idaho 67 and the Mountain Home Air Force Base road. Turn right (west) on Idaho 67, toward Grand View. *(16.65)*

25.55 In Grand View turn right (west) on Idaho 78. *(11.85)*

37.4 Castle Creek crossing. *(0.45)*

37.85 Turn right (north) on the Wees Road (gravel). *(3.6)*

41.45 Turn right (east) on the Nettleton Ranch road at the "Nettleton Ranch" sign. The large hill on the right is Castle Butte. *(1.0)*

42.45 Cross a cattle guard at the ranch gate and immediately turn left (north). *(0.4)*

42.85 Turn left (north) toward two pillars of basalt tuff. *(0.1)*

42.95 **Stop 7: Castle Butte volcanic complex.**

Points of interest: massive basaltic tuff surrounding large sediment inclusions; semi-welded, bedded, basaltic tuff.

Park near the tuff pillars. The adjacent large hill is the northwest part of the Castle Butte volcanoic complex that may contain one or more eruptive points. Walk north to the two spires or pillars. These pillars are composed of massive basaltic tuff made of fresh-appearing cinders and cindery bombs. Within the tuff are large inclusions of the underlying fine-grained sediments which have been offset by small faults in the pillar to the right. Walk west toward a cleft in the massive tuff and follow it up to the top of the hill. This cleft is an owl nesting place, and the ground and rocks are littered with owl pellets. The massive tuff exposed in the cleft contains inclusions of rounded, somewhat porcelaneous, fine-grained, white sediments (Figure 10).

At the top of the cleft, walk to the right along a ridge to a small flat-topped butte (elevation

Figure 10. Large, rounded, porcelaneous inclusion of fine-grained sediments within the massive basaltic tuff that makes up most of the northwest hill in the Castle Butte volcanic complex (Stop 7).

point 2556 feet) on the west side of the hill. You will walk across the massive basaltic tuff and an area of unconsolidated pebble gravel and sediments to a small exposure of bedded, semi-welded, cindery basaltic tuff. Because of the relation of this bedded tuff to the sediments and the massive volcanic tuff beneath it, we suggest it is a remnant of the Castle Butte area tuff, which is also exposed in the sediments beneath the basalt rim on the east side of the Snake River canyon directly across from this location (Jenks and Bonnichsen, in press(b)).

The tuff at this stop contains numerous, excellent examples of volcanic and sedimentary structures and consists of two major zones. The lower zone overlies structureless, fine-grained white sediments and consists of thinly bedded to layered sediments and basaltic ash and cinders. The most striking features within this lower zone are the numerous and complex flame, load, and rip-up clast structures within the thickest bed.

Figure 11. Large inclusion of layered fine-grained sediments within the upper zone of a remnant of the Castle Butte area tuff on top of the northwest hill of the Castle Butte volcanic complex (Stop 7). Beneath the massive semi-welded tuff that contains inclusions and dewatering channels are thinly bedded layers of cinders, sediments, and basaltic tuff.

The upper zone has a lower part that is thinly bedded and consists predominantly of fine-grained, white sediments. These lower beds have been disrupted into decollement folds and large rip-up clasts (Figure 11) by the emplacement of the overlying, more massive and structureless basaltic tuff. Within this upper layer of massive basaltic tuff are numerous vertical dewatering channels.

Return to the vehicles through the cleft. Follow the Nettleton Ranch road and the Wees Road to Idaho 78. *(5.1)*

48.05 Turn right (west) on Idaho 78. *(12.2)*

60.25 Sinker Creek crossing. The small canyon cut by Sinker Creek at this crossing exposes almost 200 feet of water-affected and pillowed basalt flows that were erupted into a Lake Idaho lake from the small hill approximately ¼ mile to the right (northeast) of the highway, Hill 3337 (elevation). The water-affected appearance of these flows is not as well developed as it is at the Hot Creek basalt at Stop 6. *(1.65)*

61.9 Pass the Silver City Road on the left. *(1.2)*

63.1 Turn right (east) on the paved road just past milepost 33. Follow the pavement east until it ends at Warrer Road, just before a cattle guard. The road climbs to and travels over the Murphy Flat that is underlain by basalt lava flows and tuffs.

In the early 1900s Murphy Flat was the site of an irrigation scheme (Nettleton, 1978). The

Murphy Land and Irrigation Company built an earthen dam upstream on Sinker Creek and diverted the water through canals onto Murphy Flat. The homesteaders on Sinker Creek contested the water rights, but the company was able to purchase sufficient land to complete the project. Six thousand acres of land were sold to prospective homesteaders, but water was available for only 800 acres because the dam leaked. At the height of the irrigation project, about 100 people lived on the flat, and a school and store were built. Within 10 years the company went bankrupt and the project was sold to other interests. On June 19, 1943, the dam burst and sent a wall of water 60 feet high down Sinker Creek, taking everything with it. Another dam was built in the 1970s, and once again water is supplied to the farming operations on Murphy Flat. *(5.1)*

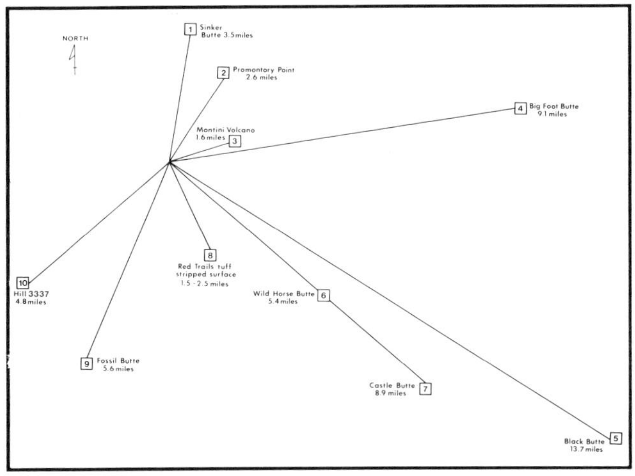

Figure 12. Panoramic index of volcanic features visible from the Sinker Creek Butte volcano (Stop 8). See text for description.

68.2 Cross the cattle guard and continue east on a gravel road. *(1.3)*

69.5 **Stop 8: Sinker Creek Butte viewpoint.**

Points of interest: Sinker Creek Butte volcano; panorama of volcanic features in the Sinker Creek area.

Park at the small turnout on the right (east) side of the road just before the Nahas Ranch sign and before the road begins its descent down a draw into Sinker Creek canyon. Walk to the left (northeast) around the north rim of the draw to the highest point on Sinker Creek Butte (VABM 3129).

Some scientific controversy exist as to whether Sinker Creek Butte is a volcano, or the result of structural deformation (Amini, 1983), but our research confirms it is a volcano. The basalt flows exposed on the top of Sinker Creek Butte have a different phenocryst assemblage than the flows that form the north rim of the draw. We have called the flows from Sinker Creek Butte volcano, the Otter Massacre Site basalt (Jenks and Bonnichsen, in press (b); Jenks, Bonnichsen, and Godchaux, in prep.). The Otter Massacre Site basalt was erupted subaerially, but it appears to have flowed into a small Lake Idaho lake basin. Thus, all of the flows were affected by water and are a combination of pillowed and water-affected facies. Probably the last eruptive products of the Sinker Creek Butte volcano were pyroclastic tuffs that may have resulted when water inundated the vent of the volcano. The Otter Massacre Site basalt thins to the south and east and is the lowest unit in a basin-filling sequence. Where it is exposed near the bottom of the canyon along the Emigrant Trail grade, the Otter Massacre Site basalt contains basalt columns that are water-affected and altered on the inside. Amini and others (1984) report two K-Ar whole rock age dates of 1.26 ± 0.17 and 6.05 ± 0.17 Ma for a sample from the top of Sinker Creek Butte volcano. Figure 12, a panoramic index of the volcanic features that can be seen from Stop 8 at the top of Sinker Creek Butte, is keyed to the following descriptions:

1. Sinker Butte is the most prominent topographic feature in the area and was used as a landmark by cowboys and by wagon trains on the nearby Emigrant Trail (Nettleton, 1978). Probably initially erupted into a Lake Idaho lake, Sinker Butte volcano has a base that is mainly hydroclastic with a thick sequence of bedded, fresh and palagonitized, basaltic tuff exposed on its north, south, and east sides. The volcano eventually grew to above the lake level and is capped at 3300 feet of elevation by a rim of subaerial flows and welded basalt spatter. The entire edifice is cut by radial dikes that are visible in the west wall of the Snake River canyon at Swan Falls Dam. In addition the lake may have been receding at the time of the last eruptions because subaerial Sinker Butte flows filled paleotopographic lows around the volcano. The flows at the north rim of the draw west of Sinker Creek Butte are Sinker Butte basalt flows. Sinker Butte basalt has a distinctive lithology of cumulophyric rosettes and clumps of large (up to 1.5 centimeters) plagioclase laths with large (up to 0.25 centimeter) olivine phenocrysts. K-Ar whole-rock age dates for the Sinker Butte basalt flows range from 0.88 ± 0.11 to 1.52 ±0.07 Ma (Amini and others, 1984).

2. Promontory Point at elevation 3008 feet in the Sinker Butte quadrangle is a headland in the

east wall of the Snake River canyon. Several of the volcanic units exposed downstream in the walls of the canyon end at Promontory Point (Jenks and Bonnichsen, in press(b)). These include (1) the Swan Falls Reservoir basalt, which is the lowest unit; (2) the Peregrine Falcon tuff, which overlies the Swan Falls Reservoir basalt and may have been erupted at least in part from the Sinker Butte volcano; and (3) the Sinker Butte basalt, which ends unconformably against a slope of Peregrine Falcon tuff. Overlying these units is the Promontory Point basalt, which also is the lower thick basalt in the east rim of the large alcove to the south of Promontory Point.

3. The crater area of Montini volcano is the large depression visible in the east rim of the Snake River canyon. Probably a maar-like volcano, the central crater depression of the volcano and a crater-filling flow are visible from this viewpoint. The central area of the volcano will be viewed at Stop 9.

4. On the horizon to the east is Big Foot Butte, a large shield volcano. Probably one of the older volcanoes in the area, Big Foot Butte volcano is surrounded by flows (Jenks and Bonnichsen, in press (b)) erupted from volcanoes to the north and east. These volcanoes, which include Christmas Mountain, Coyote Butte, Initial Point, and Wind Caves volcano form a ridge down the center of the western Snake River Plain. Their flows extend to the rim of the Snake River canyon.

5. Another landmark on the Emigrant Trail is Black Butte visible on the skyline on the east side of the Snake River canyon. Black Butte consists of dipping, subaerial-appearing, basalt flows. Jackass Butte, a hydroclastic volcano, is adjacent to Black Butte on the west side of the Snake River, but is not visible from this viewpoint.

6. Wild Horse Butte in the middle ground consists of a rim of basalt containing two basalt units are upper, Wild Horse Butte basalt and a lower Red-Legged Hawk basalt (Jenks and Bonnichsen, in press(b)), above Idaho Group sediments. Among the older units in the area with K-Ar whole-rock age dates of 1.92± 0.16 and 3.87 ± 0.28 Ma (Amini and others, 1984), the basalt units were erupted from volcanoes to the north and east. The base of the Red-Legged Hawk basalt contains areas of pillowed lava, suggesting that it flowed into a Lake Idaho lake; the top of the Red-Legged Hawk basalt as well as Wild Horse Butte basalt are subaerial in appearance. Stratigraphic relationships suggest that Wild Horse Butte and the unnamed butte northwest of it were the rim of a Lake Idaho lake basin that was filled with younger basaltic flows and tuff units including the Otter Massacre Site basalt and the Red Trails tuff, as well as fine-grained sediments (Jenks and Bonnichsen, in press(b)). This older basin rim was subsequently exhumed when the present Snake River canyon was carved.

7. Castle Butte volcanic complex (previously described at Stop 7) is visible behind Wild Horse Butte.

8. The Red Trails tuff is one of the most widespread units in the Sinker Creek area. It underlies the flat area visible below the Wild Horse Butte rim in the middle ground. Because it is part of a basin-filling sequence, the Red Trails tuff is lower in elevation but younger than the basalt units on Wild Horse Butte. Probably stripped of overlying sediments by the Bonneville Flood, the Red Trails tuff is a distinctive marker of palagonitized and fresh pyroclastic materials (Jenks and Bonnichsen, in press(b); Jenks, Bonnichsen, and Godchaux, in prep.). Stratigraphically, the Red Trails tuff is above the Otter Massacre Site basalt and below the Nahas Ranch basalt and the tuffs and flows erupted from the Montini volcanic center. Because it is thickest in the area surrounding the mouth of Sinker Creek, it is possible that this tuff was erupted in part from Sinker Creek Butte or from an earlier volcano located in the same general area. The structures within the Red Trails tuff will be viewed at Stop 10.

9. Fossil Butte, another landmark on the Emigrant Trail, is the hill to the south. From our preliminary work, this volcano appears to be similar to Sailor Cap Butte volcano because it consists mainly of a thick palagonitized tuff overlain by a cap or crater lake of subaerial-appearing basalt flows. Fossil Butte is stratigraphically above the flows from Hill 3337 (elevation). Samples of Fossil Butte basalt have been dated at 7.07 ± 0.76 and 7.24 ±0.29 Ma (Amini and others, 1984).

10. North of Fossil Butte, the Hill 3337 (elevation) volcano is visible on the skyline.

Return to the vehicles. *(1.2)*

70.7 Descend into the Sinker Creek canyon and turn left (northeast) at the junction just before the wooden Nahas Ranch gate. At this junction the road crosses the southern alternate route of the Oregon Trail called the Emigrant Trail. The grade up to the rim to the north is the site of a mid-nineteenth century Indian-pioneer confrontation called the Otter Massacre. The Otter party, which consisted of eight wagons containing four families and eight single men, had come west from Geneva, Wisconsin. Delayed for unknown reasons, they reached Sinker Creek on September 7, 1860, unaware that Fort Boise had been abandoned earlier because of Indian unrest in the area. As they approached the north rim of the canyon they were attacked and chased by a band of one hundred Indians. Eleven people in the party, including most of the Otter family, were killed outright. The rest were forced to abandon their wagons and provisions and seek refuge near the Snake River. From there, they made their way on foot to near the area of present-day Adrian, Oregon, where they erected temporary shelters. Several small groups from the remaining party left the camp to gain aid from the troops at Fort Walla Walla in Washington Territory. Finally, three men, starving and nearly naked, arrived at the fort on October 2. Soldiers from the fort rescued the remaining twelve survivors who had resorted to cannibalism of their dead in order to survive (Adams, 1979).

Go through the metal gate and drive along the north side of Sinker Creek to its mouth with the Snake River. The small fenced area on the left just past the first metal gate contains the unmarked graves of two children of the Jacob Rueben (or Rubin) family who homesteaded this side of the creek around 1880. Jacob Rueben was notorious in the Sinker Creek area because he was married to an Indian woman and because he caught wild burros and sold their meat as beef to the mining camps in the nearby Owyhee Mountains (Nettleton, 1978). *(1.9)*

72.6 **Stop 9: Montini volcano.**

Points of interest: crater area of a maar-like volcano including remnant rims, dikes, and central collapsed area.

The Montini volcano is a large, maar-like, collapse structure. It erupted a relatively small amount of juvenile tuffaceous material that only extends a maximum of 2 miles from the crater. Based on the crosscutting relationships visible in the crater rim on either side of the central depression, the tuff from the Montini volcano appears to be confined to the red pyroclastic materials (Jenks and Bonnichsen, in press (b); Jenks, Bonnichsen, and Godchaux, in prep.).

The west side of the central crater of the Montini volcano has been dissected by the Snake River, leaving only isolated pieces of the crater rim exposed on the west side of the river. In the main part of the crater on the east side of the river, the central explosively excavated and collapsed area contains jumbled blocks of Red Trails tuff, areas of massive (fresh?) basaltic tuff, and closely jointed dikes. Contained within the central depression is a flat-topped flow that probably represents the crater lava lake which was erupted after the main collapse.

On the west side of the Snake River on both sides of the mouth of Sinker Creek are more crater rim remnants. The spires on the south side of Sinker Creek are composed of a massive, basaltic, cindery tuff similar to the tuff at the Castle Butte volcanic complex at Stop 7. On the north side of the creek is a dike of closely jointed basalt that merges into a small lava flow at its top. Also exposed on the north side of Sinker Creek, but not visible from the canyon bottom, is an outcrop of welded basaltic tuff or agglutinate erupted from the Montini volcano.

A K-Ar whole rock age date of 1.64 ± 0.18 Ma has been obtained from a sample of the plug basalt of the Montini volcano (Amini and others, 1984). The basaltic tuff from the Montini volcano is stratigraphically below the Promontory Point basalt and the Sinker Butte basalt (Jenks and Bonnichsen, in press(b); Jenks, Bonnichsen, and Godchaux, in prep.). *(2.7)*

75.3 Turn left (east) at the Nahas Ranch gate junction (same junction as mileage point 70.7). Follow the road through the ranch buildings and around the northeast side of a small lake. *(2.4)*

77.7 Turn left (north) at the junction at elevation point 2421 feet in section 20 of the Wild Horse Butte 7 ½-minute quadrangle. Proceed through the gate in the fence. As you drive up the road, the edge of the Otter Massacre Site basalt is visible within the fine-grained sediments that underlie the Red Trails tuff. *(0.5)*

78.2 **Stop 10: Red Trails tuff.**

Points of interest: eruptive structures in a bedded, basaltic, pyroclastic tuff.

Numerous complex structures within a bedded basaltic tuff are well-exposed in the low cliffs on both sides of the road. The best exposures are in the cliffs to the right (east) of the road. Within the Red Trails tuff, but above the bedded pyroclastic tuff, which is the focus of this stop, is a section of massive basaltic tuff containing large pieces of basalt and basaltic cinders. This massive tuff is actually detritus from the advancing pillow delta of the Nahas Ranch basalt, which is exposed to the east and overlies the Red Trails tuff.

Figure 13. View of the Red Trails tuff in the Morrison Gulch area (Stop 10). In the center of the photograph is an area of cindery, massive matrix that encloses two large, rounded blocks of fine-grained, layered basaltic tuff.

Figure 14. Flame structures and convoluted bedding in the lower part of the Red Trails tuff at Stop 10. The largest fold is approximately 10 inches tall.

The Red Trails tuff in this area is a series of thin and thick, mostly palagonitized, basaltic tuff layers, which were deposited within a Lake Idaho lake basin. Many of the layers appear to have been disturbed during or just after their emplacement. The thickest layer in the bedded tuff contains very large, somewhat rounded blocks of fine-grained tuff that appear to be jumbled at many attitudes within a matrix con-

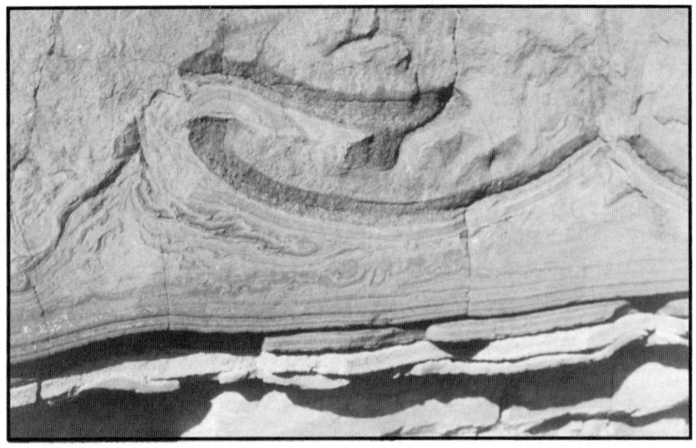

Figure 15. Surge-type cross-bedded layers in the Red Trails tuff (Stop 10).

taining fine-grained pyroclastics as well as chunks of cinders (Figure 13). Beneath the massive tuff is a layer made up of only fine-grained pyroclastics that has been deformed to extremely complex flame structures and convoluted laminae (Figure 14). Also visible in several areas below the convoluted layers are small, surge-type, crossbedded layers (Figure 15). All of these structures suggest that the bedded tuffs were disturbed by the weight or eruptive energy of pyroclastic flows from a nearby eruption.

Return to Idaho 78 via the Nahas Ranch road. *(10.65)*

88.85 Turn right (north) on Idaho 78. *(13.5)*

102.35 Turn right (north) on Idaho 45. *(17.0)*

119.35 Follow the signs in Nampa to I-84 eastbound to Boise. *(15.0)*

134.35 Boise airport exit.

ACKNOWLEDGEMENTS

For their assistance in the field, we wish to thank Dorothy and Joseph Dana, Martha Godchaux, and Gordy and Ralph Jenks. Nick and Maudy Nettleton graciously gave us permission to work on their land. We wish to thank Kurt Othberg and Bob DeTar for their thoughtful reviews of the manuscript, and the participants of the 1988 COGEOMAP field conference for their many helpful comments and suggestions. Charlotte Fullerton helped with the preparation of the manuscript. Finally, we wish to thank the University of Idaho Research Foundation for providing the funds for us to pursue in the field the concept of using basalt stratigraphy to better understand the history of the Lake Idaho lake system.

REFERENCES

Adams, Mildretta, 1979, Owyhee cattlemen, 1878-1978; 100 years in the saddle (rev. ed.): Homedale, Idaho, Owyhee Publishing Company, 194 p.

Amini, M. H., 1983, Geochronology, paleomagnetism and petrology of the upper Cenozoic Bruneau Formation in the western Snake River Plain, Idaho: University of Colorado Ph.D. dissertation, 209 p.

Amini, M. H., H. H. Mehnert, and J. D. Obradovich, 1984, K-Ar ages of late Cenozoic basalts from the western Snake River Plain, Idaho: Isochron/West, no. 41, p. 7-11.

Armstrong, R. L., W. P. Leeman, and H. E. Malde, 1975, K-Ar dating, Quaternary and Neogene volcanic rocks of the Snake River Plain, Idaho: American Journal of Science, v. 275, p. 225-251.

Armstrong, R. L., J. E. Harakal, and W. M. Neill, 1980, K-Ar dating of Snake River Plain (Idaho) volcanic rocks—new results: Isochron/West, no. 27, p. 5-10.

Bonnichsen, Bill, and M. D. Jenks, 1988, Variations in the style of basaltic volcanism, central and western Snake River Plain, Idaho (abs.): Geological Society of America Abstracts with Programs, v. 20, no. 6, p. 407.

———, in press, Geologic map of the Jarbidge River Wilderness Study Area, Owyhee County, Idaho: U.S. Geological Survey Miscellaneous Field Studies Map, scale 1:50,000.

Bonnichsen, Bill, W. P. Leeman, M. D. Jenks, and Norio Honjo, 1988, Geologic field trip guide to the central and western Snake River Plain, Idaho, emphasizing the silicic volcanic rocks, *in* P.K. Link and W.R. Hackett, editors, Guidebook to the Geology of Central and Southern Idaho: Idaho Geological Survey Bulletin 27, p. 247-281.

Buwulda, J. P., 1921, Oil and gas possibilities of eastern Oregon: Oregon Bureau of Mines and Geology, v. 3, no. 2, p. 1-47.

Cope, E. D., 1870, On the fishes of a fresh-water Tertiary in Idaho, discovered by Capt. Clarence King: Proceedings of the American Philosophical Society, v. 11, p. 538-547.

———, 1883a, On the fishes of the Recent and Pliocene lakes of the western part of the Great Basin, and of the Idaho Pliocene lake: Proceedings of the Academy of Natural Science, Philadelphia, p. 134-166.

———, 1883b, A new Pliocene formation in the Snake River valley: American Naturalist, v. 17, no. 8, p. 867-868.

Ekren, E. B., D. H. McIntyre, and E. H. Bennett, 1984, High-temperature, large-volume, lavalike ash-flow tuffs without calderas in southwestern Idaho: U.S. Geological Survey Professional Paper 1272, 76 p.

Ekren, E. B., D. H. McIntyre, E. H. Bennett, and H. E. Malde, 1981, Geologic map of Owyhee County, Idaho, west of 116 degrees west longitude: U.S. Geological Survey Miscellaneous Investigations Series Map I-1256, scale 1:125,000.

Gallegos, Dora, Peggy Johnson, Spencer Wood, and Walter Snyder, 1987, Depositional facies patterns along the Boise front, *in* Spencer Wood and J. W. Sears, editors, Field Trip Guidebook for the 1987 Rocky Mountain Section Meeting of the AAPG-SEPM: Northwest Geology, v. 16, p. 47-59.

Hart, W. K., and J. L. Aronson, 1983, K-Ar ages of rhyolite from the western Snake River Plain area, Oregon, Idaho, and Nevada: Isochron/West, no. 36, p. 17-19.

Hayden, F. V., 1872, U.S. geological survey of the Territories: Fourth Annual Report, 2nd Annual Preliminary Report (of Wyoming and portions of contiguous territories), v. 4, no. 4, 511 p.

Jenks, M. D., 1984, A morphometric analysis and classification of the Neogene basalt volcanoes in the south-central Snake River Plain, Idaho: University of Idaho M.S. thesis, 131 p.

Jenks, M. D., and Bill Bonnichsen, 1987, Lake Idaho; new perspectives through basalt stratigraphy (abs.): American Association of Petroleum Geologists, v. 71, p. 1008.

———, in press(a), Geologic map of the Bruneau River Wilderness Study Area, Owyhee County, Idaho: U.S. Geological Survey Miscellaneous Field Studies Map, scale 1:50,000.

———, in press(b), Geologic map of the Wild Horse Butte quadrangle, Ada and Owyhee Counties, Idaho: Idaho Geological Survey Technical Report, scale 1:24,000.

Jones, F. G., 1978, The Hot Spring Limestone – Tertiary algal reefs from the Snake River Plain, Idaho: University of Michigan M.S. thesis, 85 p.

Kimmel, P. G., 1982, Miocene-Pliocene lacustrine sediments of the western Snake River Plain: their stratigraphy, age, and tectonic setting, *in* Bill Bonnichsen and R. M. Breckenridge, editors, Cenozoic Geology of Idaho: Idaho Bureau of Mines and Geology Bulletin 26, p. 559-578.

King, Clarence, 1878, Systematic geology: U.S. Geological Exploration of the 40th parallel, v. 1, Washington, D.C., 803 p.

———, 1903, Mountaineering in the Sierra Nevada: New York, Scribners, 378 p.

Kirkham, V. R. D., 1931, Revision of the Payette and Idaho Formations: Journal of Geology, v. 39, no. 3, p. 193-239.

Lindgren, Waldemar, and N. F. Drake, 1904, Description of the Nampa quadrangle: U. S. Geological Survey, Geologic Atlas Folio 103, scale 1:125,000.

Malde, H. E., 1972, Stratigraphy of the Glenns Ferry Formation from Hammett to Hagerman, Idaho: U.S. Geological Survey Bulletin 1331-D, p. 1-19.

Malde, H. E., and H. A. Powers, 1962, Upper Cenozoic stratigraphy of western Snake River Plain, Idaho: Geological Society of America Bulletin, v. 73, p. 1197-1220.

Malde, H. E., H. A. Powers, and C. E. Marshall, 1963, Reconnaissance geologic map of west-central Snake River Plain, Idaho: U.S. Geological Survey Miscellaneous Geologic Investigations Map I-373, scale 1:125,000.

Middleton, L. T., M. L. Porter, and P. G. Kimmel, 1985, Depositional settings of the Chalk Hills and Glenns Ferry Formations west of Bruneau, Idaho, *in* R. M. Flores and S. S. Kaplan, editors, Cenozoic Paleogeography of West-central United States: Society of Economic Petrologists and Mineralogists, Rocky Mountain Section, v. 8, p. 37-53.

Miller, R. R., and G. R. Smith, 1967, New fossil fishes from Plio-Pleistocene Lake Idaho: University of Michigan Museum of Zoology Occasional Papers, v. 654, p. 1-24.

Nettleton, Helen, 1978, Sketches of Owyhee County: Nampa, Idaho, Schwartz Printing Company, 275 p.

Russell, I. C., 1902, Geology and water resources of the Snake River plains of Idaho: U.S. Geological Survey Bulletin 199, 192 p.

Smith, G. R., 1975, Fishes of the Pliocene Glenns Ferry Formation, southwest Idaho: University of Michigan Museum of Paleontology Papers on Paleontology, v. 14, p. 1-68.

Smith, G. R., Krystyna Swirydczuk, P. G. Kimmel, and B. H. Wilkinson, 1982, Fish biostratigraphy of late Miocene to Pleistocene sediments of the western Snake River Plain, Idaho, *in* Bill Bonnichsen and R. M. Breckenridge, editors, Cenozoic Geology of Idaho: Idaho Bureau of Mines and Geology Bulletin 26, p. 519-541.

Stearns, H. T., L. Crandall, and W. G. Steward, 1938, Geology and ground water resources of the Snake River Plain in southeastern Idaho: U.S. Geological Survey Water Supply Paper 774, 268 p.

Swirydczuk, Krystyna, B. H. Wilkinson, and G. R. Smith, 1979, The Pliocene Glenns Ferry oolite; lake margin carbonate deposition in the south-western Snake River Plain: Journal of Sedimentary Petrology, v. 49, p. 995-1004.

———, 1980, The Pliocene Glenns Ferry oolite: sedimentology of oolitic lacustrine terrace deposits: Journal of Sedimentary Petrology, v. 50, p. 1237-1248.

Swirydczuk, Krystyna, G. P. Larson, and G. R. Smith, 1982, Volcanic ash beds as stratigraphic markers in the Glenns Ferry and Chalk Hills Formations from Adrian, Oregon, to Bruneau, Idaho, *in* Bill Bonnichsen and R. M. Breckenridge, editors, Cenozoic Geology of Idaho: Idaho Bureau of Mines and Geology Bulletin 26, p. 543-558.

The Miocene Clarkia Fossil Area of Northern Idaho

C. J. Smiley [1]

INTRODUCTION

This field trip travels to the fossil area of Miocene Clarkia Lake in the valley of the present St. Maries River. For the 1989 Geological Society of America meeting, the field trip starts and ends in Spokane, Washington. The itinerary follows along the western side of Coeur d'Alene Lake and then southeastward to the St. Maries River and the small townsite of Clarkia, Idaho. The round-trip distance totals about 230 miles and covers examples of (1) different units of the Precambrian Belt Supergroup and older rocks that comprise the basement complex of the region; (2) Miocene basalts exposed largely at lower elevations; and (3) modern drainage systems and modern bottomland sedimentation that are considered to be comparable to the Miocene setting of the Clarkia fossil beds.

In the roadlog the itinerary (Figure 1) has been divided into several segments for the convenience of those who may not be travelling from Spokane, or who may wish to take additional side trips. The mileage notations for specific points of interest are based on odometer readings beginning at 0.0 for the start of each segment. A segment is introduced by a general statement of geology, followed by odometer readings at specific points of interest. The itinerary is divided into these segments:

Segment	Route
A	Spokane, Washington, to Coeur d'Alene, Idaho (31 miles).
B	Coeur d'Alene to Plummer (34 miles).
C	Plummer to St. Maries (19 miles).
D	St. Maries to Santa (15 miles).
E	Santa to Clarkia Site P-33 (18 miles).
F	St. Maries (via Coeur d'Alene River) to I-90 (33 miles).
G	Junction Idaho 3 and I-90 to Coeur d'Alene (22 miles).

[1] *Department of Geology and Geological Engineering, University of Idaho, Moscow, Idaho 83843.*

Smiley, C. J., 1989, The Miocene Clarkia fossil area of northern Idaho, *in* V. E. Chamberlain, R. M. Breckenridge, and Bill Bonnichsen, editors, Guidebook to the Geology of Northern and Western Idaho and Surrounding Area: Idaho Geological Survey Bulletin 28, p. 35-48.

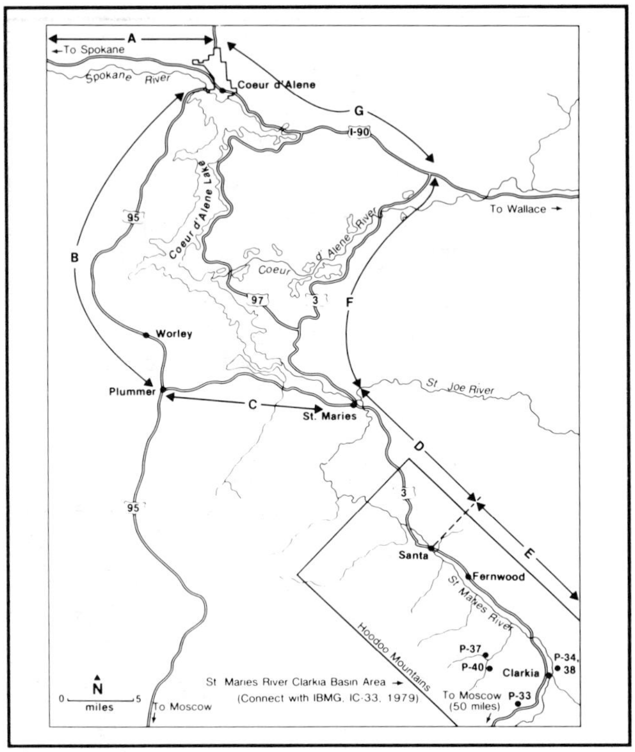

Figure 1. Map showing itinerary from Spokane, Washington, to the Clarkia basin, Idaho. The itinerary is divided into segments A-G. For roadlog reference see text.

Rock Units Along the Itinerary

The following geologic column lists the rock units mentioned in the road log.

- Pleistocene
  - Glacial and flood deposits
  - Palouse Formation loess deposits
- Miocene
  - Columbia River Basalt Group
    - Wanapum Formation
    - Grande Ronde Formation
    - Latah Formation (fossiliferous interbeds)
      - Clarkia Lake deposits
- Jurassic Cretaceous and Tertiary
  - Intrusive rocks
- Precambrian
  - Belt Supergroup
    - Libby Formation
    - Striped Peak Formation
    - Wallace Formation
    - St. Regis Formation
    - Revett Formation
    - Burke Formation
    - Prichard Formation
  - Pre-Belt metamorphic rocks

General Vegetation

Vegetation in this region reflects a continental climate of "summer-dry" and "winter-wet" extremes modified by orographic factors. Near Heyburn State Park at the southern end of Coeur d'Alene Lake is an area containing excellent examples of the plants of the region (see Segment C). Here, in a small area, can be seen (1) bog and lake-border vegetation of aquatic and other mesic plants; (2) the dicot vegetation of open sites such as forest borders, riparian habitats, and roadsides; and (3) many of the conifer species that typify the slopes and uplands of the surrounding mountains. Woody plants occurring here include the following conifers and dicots: conifers—Ponderosa pine, Idaho white pine, lodgepole pine, douglas fir, grand fir, tamarack, spruce, hemlock, and cedar; dicots—aspen, poplar, willow, alder, mountain maple, cascara, mock orange, elderberry, rose, cherry, thimble berry, service berry, snow berry, and holodiscus.

Regional and Local Climatic Factors

The present climate of the interior Pacific Northwest, described as summer-dry continental, is reflected by a diversity of vegetation types that are expressed as plant responses to warm season water-stress and cold season frost-stress. The local response to these water and temperature stresses is in large part a response to local and regional orographic factors (Daubenmire, 1969). For example, in this northern Idaho area along the western slopes of the northern Rocky Mountains, Neogene climates and vegetation are in large part the result of (1) the historical uplift of the Cascadian climatic barrier on the west, with progressive development of a rain shadow in its lee during later Neogene time and (2) orographic factors of increasing elevation and topographic diversity toward the east. For example, over the Columbia Basin to the west and southwest of this field trip area at elevations of about 1000-1500 feet and in the rain shadow of the Cascade Range, the average annual precipitation is within the range of 6-10 inches. Warm-season relative humidity typically is 10-20 percent, during the middle of the day, and cooling cloud cover is rare. In contrast, along the western slopes of the northern Rocky Mountains in northern Idaho, at elevations from 2500 to 5000 feet, the average annual precipitation ranges from 25 to 50 inches depending on elevation; in addition, montane cloud cover is common, and midday relative humidity typically is 30-50 percent (Kincer, 1941). Comparable values of temperature changes (related to elevation) also are evident, as are diurnal variations that affect both the temperature-stress and the water-stress of plants.

CLARKIA BASIN

Geologic setting of Miocene Clarkia Lake

The valley of the St. Maries River trends northwest-southeast in conformance with early Cenozoic structural lineations of the region (Griggs, 1973; Smiley and Rember, 1979, 1985a; Bennett, 1986). Running parallel to the valley to the southwest is the granite-cored ridge of the Hoodoo Mountains (Figure 2). The St. Maries River valley and the major tributaries that drain the eastern slopes of the Hoodoo Mountains are deeply incised into the Precambrian Belt metasediments. The incised drainage system suggests a direction of Miocene stream flow to the northwest, as at present, and the location of a lava dam at the northwest (down gradient) end of the basin.

The geologic history of the area is outlined as follows: (1) Precambrian deposition and subsequent metamorphism of sediments of the Belt Supergroup, resulting in the mica schist and quartzite basement of the St. Maries River area; (2) granitic intrusions in the Hoodoo Mountains probably during Cretaceous time; (3) early Cenozoic tectonism producing northwest-southeast trending faults and causing the deeply incised St. Maries River drainage system; (4) fissure eruptions of Miocene basalts to the west and southwest, with lava-damming of the St. Maries River valley on the northwest to produce Miocene Clarkia Lake; (5) continued (post-Clarkia Lake) extrusions of basaltic lavas and local pyroclastics, and later deposition of a sequence of valley sands and gravels whose precise ages are yet to be determined; (6) Pleistocene glaciation in the Rocky Mountain uplands to the east and in the Spokane-Coeur d'Alene area to the north, without any evidence of glaciation in the St. Maries River valley; and (7) erosion of post-Clarkia valley-fill stripping the unconsolidated younger sediments from the valley floor, revealing in places the upper level of the fossiliferous Clarkia clays (notably in areas A, B, and C on Figure 2).

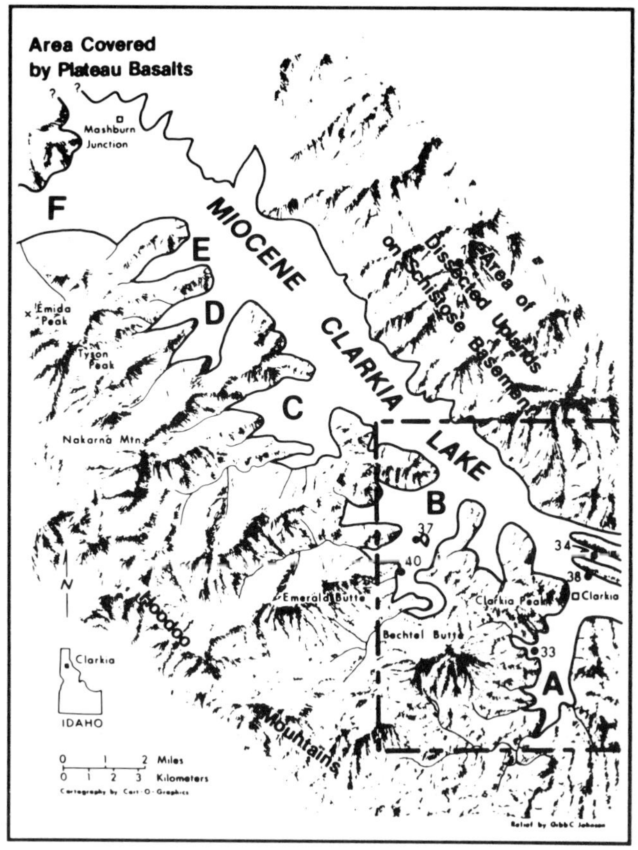

Figure 2. Inferred extent and configuration of Miocene Clarkia Lake in the valley of the St. Maries River, based on known distribution of fossiliferous lake deposits that conform to local topographic history (from Smiley and Rember, 1985a). The present trend of the St. Maries River valley conforms to early Cenozoic structural lineations (Bennett, 1986). Precambrian metasediments of the lower Wallace Formation crop out on the northeast and rocks of the upper Wallace Formation on the southwest. Areas A-F are considered to represent lateral embayments of Miocene Clarkia Lake.

The P-33 Lacustrine Section

A 9- to 10-meter-thick section of typical lake deposits is exposed at the Clarkia P-33 site (Smiley and Rember, 1979, 1985a). This appears to represent a small bay on the western (windward) side of Miocene Clarkia Lake (Figures 2, 3, 4, and 5). The sediments are soft, laminated, silty clays and volcanic ashes; they are discretely layered and unoxidized except near the top of the section (see Figure 6). Most of the local section is composed of compacted varvelike laminations, 2 to 3 millimeters thick, with interspersed clay beds 5 to 10 centimeters thick. Scattered through the section are discrete units of volcanic ash ranging in thickness from a fraction of a centimeter to 60 centimeters.

All layers show a consistent pattern of textural gradation (upward fining) with notable exceptions: (1) the thicker layers of clay commonly display internally churned features of lacustrine turbidites; (2) the thicker volcanic ash beds (10, 50 and 60 centimeters thick) contain several sequential zones of upward fining within each unit; (3) the wedgelike deposit of churned sandy clay underlies the laminated lake deposits, which may represent the slump of slope soils at the abrupt initiation of lacustrine conditions here; and (4) oxidized and poorly laminated (perhaps bioturbated) silty clays are at the top of the P-33 section.

The combined evidence of sediments and fossils here suggests rapid rates of sedimentary accumulation and fossil burial in this part of the Miocene Clarkia basin. At

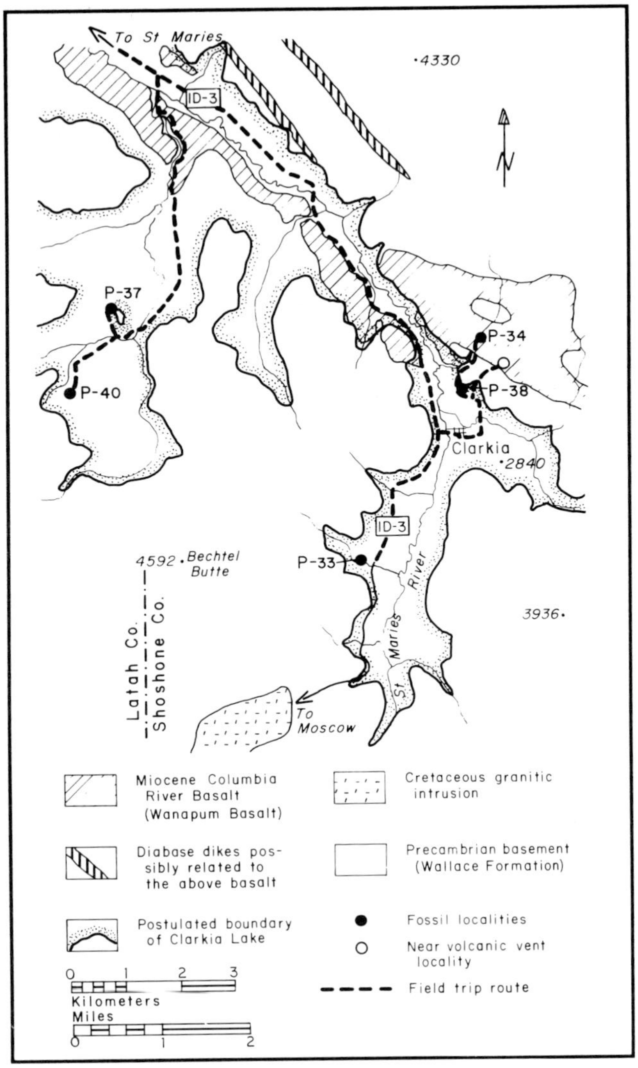

Figure 3. Local map of the updrainage end of Miocene Clarkia Lake, in areas A and B on Figure 2. The numbers 33, 34, 37, 38, and 40 are fossil sites with taxa representing a typical biota of the Miocene Clarkia deposits. In the area of sites 34 and 38 occurs a local pile of basalt flows associated with near-vent pyroclastics, underlain by fossiliferous Clarkia lake deposits that have been intruded by a dike-sill complex of the same type of basalt and blackened apparently by heat.

other sites in the basin, where megafossil and microfossil assemblages suggest an age approximately equivalent to that of the P-33 site, sedimentary environments are somewhat different in relation to water depth and to proximity to lake borders.

Megafossil Taphonomy

Dominant megafossils in the fossil-rich deposits of the Miocene Clarkia basin are plants of lake-border swamps and adjacent slopes. The result is the preservation of a

Figure 5. Four photographs show the methods of fossil collecting at the P-33 site. (A) Use of a bulldozer to break out large blocks of fossiliferous clays at the base of the exposure; (B) Hand-tool sampling in the middle of the exposure; (C) Use of the pulaski hand-tool; W.C. Rember is pointing to the contact between an unfossiliferous unit and overlying fossiliferous clay units; and (D) Search for fossils by C. J. Smiley and Francis Kienbaum in blocks broken out by a bulldozer. Photographs courtesy of Maynard M. Miller.

Figure 4. A composite of three photographs taken from a single point in the St. Maries River valley near Clarkia (from Smiley and Rember, 1985a). The panorama shows the topographic relations between the horizontal early Neogene deposits in the valley bottoms and the surrounding hills of Precambrian rocks. The P-33 exposure on the right appears to represent an uplifted and tilted fault block that has elevated the lacustrine section above its original level in the valley bottom.

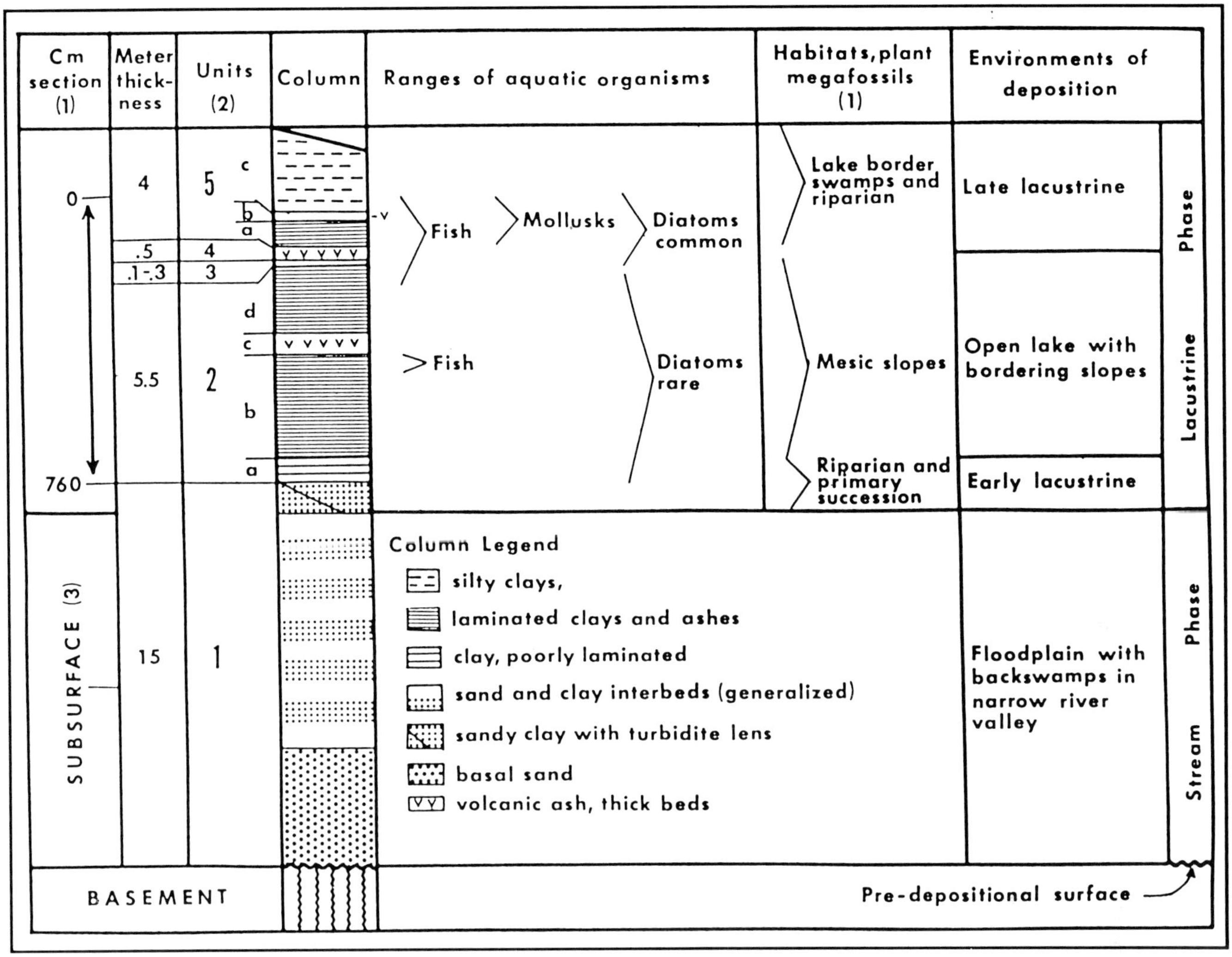

Figure 6. Stratigraphic column of the P-33 site. The basal (largely subsurface) portion is composed of alternating sands and silty clays, probably representing pre-lacustrine valley deposition (generalized from a 2.5 cm drill core at this site). The top of Unit 1 (base of the P-33 exposure) is a lens of churned sandy clay that may represent an initial slump wedge at the beginning of lacustrine conditions here. Units 2-5 show the cross-section of an essentially complete cycle of lacustrine deposition. Footnotes: (1) Smiley and Rember, 1979. (2) Smiley and others, 1975. (3) General lithology from 2.5 cm drill core.

taxonomically diverse mixed mesophytic forest assemblage in lake deposits (see Figure 7 and Plates 1 and 2 illustrating the dominant plant taxa). About forty taxa of insects represent a diverse fauna, derived largely from wet forest floor and lake border habitats. Freshwater mollusks, present in uppermost beds of the P-33 section, are rare and are represented by a species of pelecypod (a mussel) and two gastropod species. Vertebrates are represented only by fish, which are common as fossils but consist of only three or four species: the dominant sunfish (Centrarchidae, see Figure 8), the common minnowfish (Cyprinidae), and the rare large trout (Salmonidae).

Clarkia plant megafossils are remarkably well-preserved for fossils of Miocene age (Figure 7; Plates 1 and 2). They occur in taxonomic diversity and abundance, representing plants from a variety of habitats. The fine details of leaves and other plant megafossils show little or no evidence of physical abrasion from surface transport or of biological decomposition as would occur on a wet forest floor, prior to burial. Commonly they are preserved as intact cellular tissue, with intracellular structures preserved in some specimens. Many of the leaf fossils have preserved original organic chemical compounds in the cellular tissue, including those of pigmentation (reds, greens, browns). Furthermore, the distinct separation of leaves on lacustrine bedding surfaces suggests an offshore depositional site, rather than the stacking of leaf-upon-leaf that would reflect shoreline or paludal preservation.

Figure 7. Photograph of a single bedding surface (measuring 30 x 45 cm) at the P-33 site (Figure 6, base of Unit 5). The preservation of plant megafossils and their distinct separation on the bedding surface suggest an off-shore site of deposition and burial. Their taphonomy suggests removal from parent plants and short-distance (short-duration) transport by storm winds (Smiley and Rember, 1979, 1985a). A: *Alnus;* C: ?*Cocculus;* Cd: *Cercidiphyllum;* L: *Liquidambar;* Le: legume leaflet (?*Zenia);* Li: *Liriodendron;* M: *Metasequoia;* N: *Nyssa;* P: *Pseudofagus* (extinct genus); Q: *Quercus;* S: *Salix:* T: *Taxodium.*

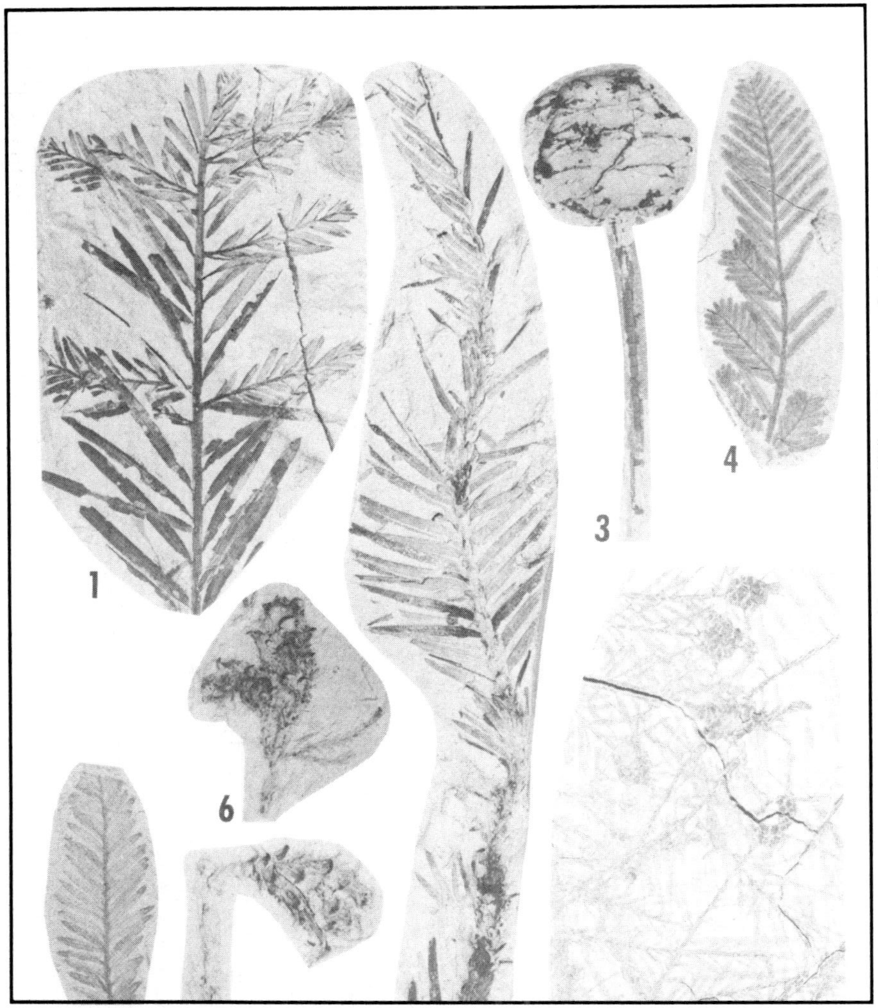

Plate 1. Common Clarkia conifers of the family Taxodiaceae. Figure 1: A leafy shoot of *Taxodium* with young side branches, Figure 2: A shoot of *Cunninghamia* (now restricted to eastern Asia), showing 3 to 4 seasons of growth. Figure 3: A compacted cone of *Metasequoia* (now restricted to eastern Asia), Figure 4: A leafy shoot of *Metasequoia* with young side branches. Figure 5: A shoot tentatively identified as *Sequoia* (now restricted to coastal habitats of western U.S). Figures 6 and 7: Shoots with attached cones of *Glyptostrobus* (now restricted to coastal-plain habitats of eastern Asia). Figure 8: A cluster of *Taxodium* shoots with insect (Dipterid) galls similar to ones on shoots of the living species in southeastern U.S. From SMiley and Rember, 1985b.

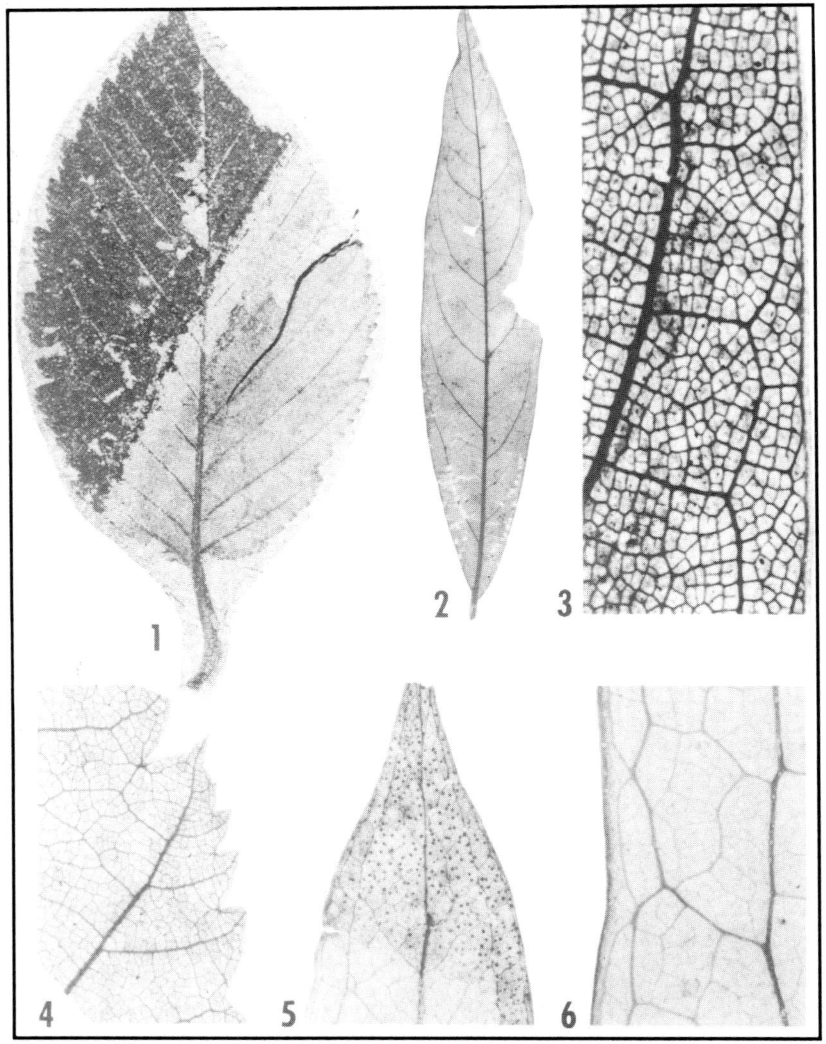

Plate 2. Examples of leaf preservation of some common Clarkia angiosperms. Figures 1 and 4: A species of birch (*Betula*), showing a partly oxidized leaf (Figure 1) and the enlarged leaf margin of a cleared compression (cellular) fossil. Figures 2 and 3: A member of the family Lauraceae (probably *Persea*) showing a cellular compression fossil (Figure 2) and the enlarged marginal portion of a cleared leaf. Figures 5 and 6: Leaf of a spinose woody vine of the genus *Smilax*, showing a fungal-infested apical portion (Figure 5) and the enlarged marginal portion of a cleared leaf. From Smiley and Rember, 1985b.

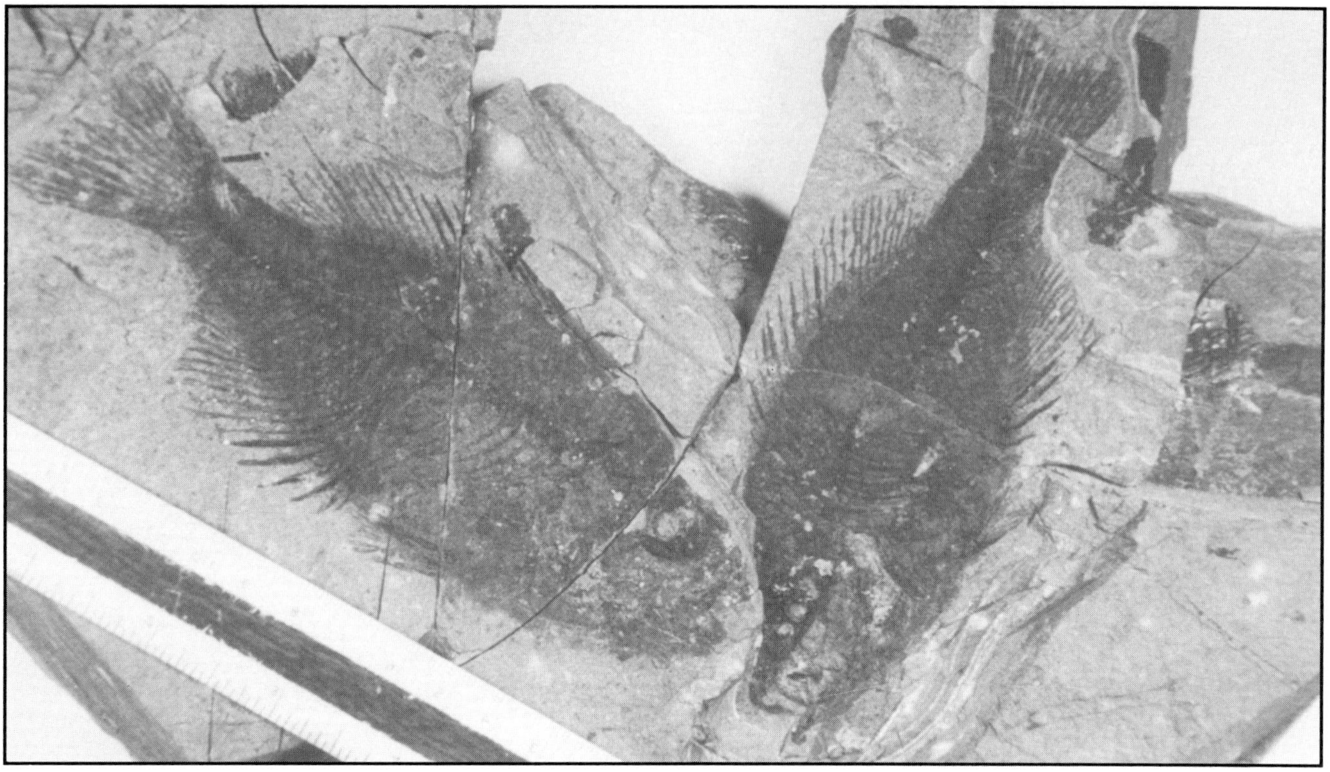

Figure 8. Two intact specimens of Clarkia fish, representing the dominant Clarkia fish *Arctoplites clarki* Smith and Miller (1985). The two fossils occur on different bedding surfaces, 1 centimeter apart, yet show the same taphonomy. A displayed tetany (open mouths and upright fins) suggests death by oxygen deficiency; what appears to be dual eyes probably is the result of the displacement of eye lenses by weak currents on the lake bottom (Smith and Elder, 1985).

Fossil insects are preserved as intact bodies or isolated body parts and as small exoskeletal fragments. Some freshly exposed specimens may show vivid original colorations similar to leaf fossils, but they also blacken rapidly upon release from the pressurized unoxidized confines of the water-saturated lake deposits. Some exoskeletal fragments are concentrated in elongate masses suggestive of the coprolites of fish. Diverse types of aquatic larval cases of caddis-flies occur in shallow-water deposits in the Clarkia basin. Insect trace fossils, represented by the mining and galling on fossil leaves and shoots, are of common occurrence.

Virtually all of the fish specimens are preserved as intact carcasses that have been flattened by compaction from the weight of overlying sediments. The surfaces of some of the specimens have been blackened from carbon-coating, and others have a greenish coating from the incrustation of pyritic material. All fish specimens display a degree of muscular tetany that has resulted in open mouths and rigidly upright fins, a taphonomy that results from death by asphyxiation. Fish bodies commonly show a lateral line of the division of body musculature, and some may display the pattern of scale distribution . Also present are isolated scales and what seem to represent fish coprolites containing insect fragments.

A few molluscan fossils are present in the uppermost beds of the P-33 section, introduced during the late stage of sedimentary infilling here. They are preserved as a single valve imprint of a pelecypod, a few coiled shells of two gastropod species, and isolated gastropod opercula. As these benthonic mollusks could not have lived in the anoxic muddy-bottom environment of the P-33 depositional site, they may have been washed out from nearby shoreline habitats to the deeper water depositional site of P-33 (Smiley, 1985, p. 73-74).

The preservation of plant and animal megafossils is restricted largely to the portion of the P-33 section that is characterized by varvelike layering. Such megafossils are not found within layers of clays and ash thicker than about 3 millimeters, except near the bottoms and tops of these rapidly deposited units. Of particular significance is the preservation of individual leaves and leafy branches crossing bedding planes, with no noticeable difference of taphonomy on the lower and higher bedding surfaces. This suggests rapid rates of sedimentary deposition in the lacustrine environment and rapid rates of fossil burial. The taphonomy of megafossils and the lack of bioturbation of organic-rich and finely laminated lake-bottom muds are inferred to signify a lacustrine depositional setting that was (1) sufficiently off-shore to result in the fine muddy texture of the P-33 deposits and (2) in sufficiently deep water to result in a permanently anoxic (perhaps toxic) hypolimnion of deposition and preservation.

Microfossil Assemblages

Microfossils of both terrestrial and lacustrine habitats are abundant and taxonomically diverse in the Miocene Clarkia deposits (various papers in Smiley, 1985). Spores and pollen are abundant and well-preserved, representing about 50 taxa that typically occur as megafossils. A diverse fungal assemblage is represented by about 200 microstructures and as epiphylls on compressed leaf megafossils (Smiley, 1985, p. 241-244). Microfossils from organisms that inhabited sites of lake-border marshes include 14 spicule types of freshwater sponges and 8 to 10 scale types of thecamoebians. Planktonic microfossils include about 50 types of diatoms, 15 cyst types of chrysophyte algae, and abundant microfossils identified as possible freshwater dinoflagellates. All of these microfossils support the evidence of megafossils and sedimentary features for paleoecological inferences.

Overview of Paleolimnology

The evidence of sedimentology, taphonomy, and assemblage analyses at the P-33 site has been combined to produce the following integrated analysis of this Miocene lacustrine system (Smiley, 1985, p. 415-417; Smiley, in press): A lake was formed abruptly by early Neogene lava-damming of a pre-existing drainage system. This produced a long, narrow, irregular body of fresh water in a hilly terrane, a lake whose surface was geographically bounded and "protected" by bordering uplands of moderate elevation. Micaceous debris derived from residual soils on adjacent slopes of the metamorphic basement, supplemented by frequent airfall deposition of volcanic ash, were rapidly deposited in the Miocene lake basin. The surrounding hills were densely clothed by a luxuriant, mixed mesophytic forest of exotic plants, reflecting a climate of humid warm-temperate conditions similar to those of the present in southeastern United States and eastern Asia (short mild winters and long wet summers). Such a climatic regime is indicated also by the Clarkia fish, insects, mollusks, the rich fungal component of the forest vegetation, and the aquatic microfossils. The subsequent change in regional climate, to the continental extremes of the present, resulted from a cooling of global climates and from later Neogene uplift of the Cascadian climatic barrier on the west.

Evidence from the P-33 site suggests the following paleolimnologic inferences for this small bay on the western (windward) side of the Miocene lake: The original depth of water must have been greater than the 9- to 10-meter thickness of compacted lake sediments that subsequently accumulated here. Compaction ratios are estimated at about 3:1 for the total P-33 column, which suggests a water depth of 20 to 25 meters at the initiation of lacustrine conditions. The water column appears to have been permanently stratified, with an epilimnion of oxygenated water of near-neutral pH to a depth of about 8 to 9 meters and an underlying denser hypolimnion of anoxic (perphaps toxic) water downward to the lake bottom. This is suggestive of a meromictic type of lake.

Sedimentary infilling progressed rapidly under the influence of downslope transport of residual soils in a humid (probably stormy) climate, and of persistent air-fall deposition of volcanic ash. The sedimentation rates in modern reservoirs and the paleontologic data suggest a probable duration for Miocene Clarkia Lake of not more than a few hundred years. During this short span of geologic time, the plant megafossil representation in the P-33 section changed markedly from bottom to top: (1) largely riparian and swamp species at the base of the section as impounded water began to flood the valley floor; (2) largely slope species in the middle of the section when the continuing impoundment of water formed shorelines higher on the valley walls; and (3) largely swamp species at the top of the section, as lake border swamps expanded outward during the later stages of lacustrine infilling. A nearly complete, progressive lacustrine cycle appears to be represented at the P-33 depositional site. Evidence supporting these inferences of Clarkia paleoecology and paleolimnology is illustrated in Figures 2-8 and by Plates 1 and 2.

ROAD LOG

Segment A: Spokane to Coeur d'Alene (31 miles)

Start at east edge of Spokane on Interstate 90 (I-90) and end at the junction of U.S. Highway 95 (U.S. 95) in Coeur d'Alene. The route travels eastward along the valley of the Spokane River, a valley that served as the "sluice-way" for Pleistocene floodwaters from glacial Lake Missoula in Montana during collapses of ice dams. Such ice-dam failures appear to have occurred repeatedly in late Pleistocene time. To the southwest of Spokane, the thick loess deposits of the Pleistocene Palouse Formation were stripped away locally, as were the underlying flows of Miocene basalts, to form the famous Channeled Scabland of eastern Washington (Griggs, 1973). Sand and gravel deposits from this glacial activity can be seen in many of the roadcuts between Spokane and Coeur d'Alene, and they also formed a dam resulting in the present Coeur d'Alene Lake.

Occasional small outcrops of Miocene basalt can be seen on the lower slopes of mountains bordering the

Spokane River valley. Southward from Coeur d'Alene, at lower elevations along both sides of Coeur d'Alene Lake, this basalt is also present and extends for some distance up tributary valleys.

The forested hills on the north and south of the Spokane River valley are composed of Precambrian metamorphic rock of the Belt Supergroup (Prichard Formation) according to Griggs (1973), but referred to as Pre-Belt metamorphic rock by Bickford and others (1985). These rocks are intruded by Mesozoic and Cenozoic granite, according to Miller and Engles (1975). The geologic history here may be summarized under five headings: (1) Precambrian basement; (2) Mesozoic and Cenozoic granite intrusions; (3) Paleogene development of drainage systems; (4) Miocene basalts largely in bottomlands; and (5) Pleistocene glaciation, loess deposition, erosion, and flood-sediment deposition.

Mileage	Description
0.0	First exposure of Miocene basalt at east edge of Spokane along I-90. Ahead is the broad, flat valley of the Spokane River, floored by glacial flood sands and gravels (Pleistocene).
25.7	Rathdrum Prairie to the north (left) is a valley carved by a lobe of the Cordilleran Ice Sheet from Canada.
31.0	Intersection of I-90 and U.S. 95 at Coeur d'-Alene, Idaho.

Segment B: Coeur d'Alene to Plummer (34 miles)

At the junction of I-90 and U.S. 95, turn south (right) on U.S. 95 and travel to Plummer, Idaho, which is at the junction of U.S. 95 and Idaho Highway 5. This segment of the itinerary follows along the west side of Coeur d'Alene Lake at a distance of 3-5 miles from the lake border. General geology is based largely on Griggs (1973). The itinerary progresses southward along U.S. 95 and traverses Precambrian basement rocks. Local patches of Columbia River basalt can be seen mostly at lower elevations in valley bottoms. Quaternary deposits (loess and valley-fill) are also present.

Mileage	Description
0.0	Junction of I-90 and U.S. 95 in Coeur d'Alene.
1.0	Bridge across the Spokane River at the outlet of Coeur d'Alene Lake. The flat bottomland is the result of Quaternary filling. The hill directly ahead is Columbia River basalt. Surrounding forested uplands are composed of metamorphic rocks of the Belt Supergroup, mapped by Griggs (1973) as the Prichard Formation.
2.7	Roadcuts are through Prichard Formation and biotite gneiss.
3.9	Cross Cougar Creek. The flat valley is formed by Quaternary filling of the lower course of a tributary valley of Coeur d'Alene Lake. The adjacent slopes and hills are Precambrian Prichard Formation and pre-Belt metamorphic rocks.
7.9	Miocene basalt outcrops (identified as Grande Ronde and upper Wanapum types) cap deeply weathered pre-Belt gneiss.
9.2	Cross Mica Creek. Exposures from here to Fighting Creek are Prichard, according to Griggs (1973), but they may also include pre-Belt high-grade metamorphic rocks.
14.3	Intersection with Rockford Bay road, on the left. Continue on U.S. 95.
15.3	Cross Fighting Creek.
19.6	Cross Lake Creek. Exposures of Wanapum Formation (Miocene basalts).
24.2	Junction of U.S. 95 and Idaho Highway 58. Open, rolling farmlands are on the Pleistocene loess deposits of the Palouse Formation. Continue on U.S. 95.
28.0	Town of Worley. Surrounding forested hills are Belt metasediments, mapped by Griggs (1973) as Revett and Burke Formations (lower-middle Belt).
32.7	Benewah-Kootenai County line.
34.3	Town of Plummer. Junction of U.S. 95 and Idaho 5. Turn east (left) on Idaho Highway 5 to St. Maries. Forested hills to south and southeast are mapped as Striped Peak Formation metamorphics (upper Belt). Miocene basalts crop out in the area as exposures in stream valleys.

Segment C: Plummer to St. Maries (19 miles)

At junction of U.S. 95 and Idaho 5, turn east (left) to St. Maries. The itinerary crosses the southern end of Lake Coeur d'Alene to the valley of the St. Joe River and to the junction with the St. Maries River. In this area the basement rocks are Belt metasediments, and Miocene basalts are exposed as a flow or a sequence of flows in valley bottoms and along lake borders. The forested uplands north of the St. Joe River valley are older Belt (Prichard Formation) whereas those to the south are younger Belt (Striped Peak Formation), indicating that the St. Joe River follows a major fault line. The mouth of the St. Joe River, where it enters the south end of Coeur d'Alene Lake, is extended northward across the lake between natural levees. Eastward, the broad river valley contains the horizontal valley deposits abutting directly against Precambrian Belt metasediments (similar to the topography inferred as the setting of Miocene Clarkia Lake in the St. Maries River valley to the southeast).

Mileage	Description
0.0	Plummer. Junction of U.S. 95 and Idaho 5.
2.2	Wanapum Basalt (Priest Rapids Member) is exposed in the valley of Little Plummer Creek. Basement rocks of the area are Striped Peak Formation.
5.0	Road descends through columnar basalts. Grande Ronde and Wanapum (Dodge flow) basalts have been identified here.
6.8	Heyburn State Park, south end of Coeur d'Alene Lake. Here are excellent examples of plant species and vegetation types mentioned in the *Introduction*.
8.3	Cross Cottonwood Creek. The view to the north (left) overlooks the south end of Coeur d'Alene Lake, where the mouth of the St. Joe River extends northward across the lake between natural levees.
9.3	Exposures of Striped Peak Formation on the right, levees of the St. Joe River on the left.
10.6	A part of Coeur d'Alene Lake east of the St. Joe River levees. Rapid Quaternary filling and the formation of bottomland bogs are evident. The flat valley floor abuts directly against metamorphic rocks of valley walls here.
11.8	Benewah Lake junction and Miocene basalt exposures.
12.4	Exposures of light-colored quartzite of the Striped Peak Formation.
13.2	High viewpoint overlooking the valley of the St. Joe River. The town of St. Maries can be seen in the distance. Older Belt rocks (Prichard) are north of the valley, and younger Belt rocks are on the south (Striped Peak and Wallace Formations). The road descends to the valley floor through Striped Peak rocks.
17.7	West edge of St. Maries. The quarry on the right is Columbia River basalt.
19.2	Junction of Idaho 5 and Idaho Highway 3 on the east side of downtown St. Maries. The itinerary continues east on Idaho 3 (straight ahead), then turns south toward Clarkia just outside of town.

Segment D: St. Maries to Santa (15 miles)

The area traversed in this segment extends from St. Maries in the valley of the St. Joe River to the vicinity of Santa near the northwest end of the Miocene Clarkia basin in the valley of the St. Maries River. From St. Maries, the road ascends a plateau on the top of an essentially horizontal sequence of Wanapum Formation and then descends through the same basalt sequence on the south. Near the base of the south side of this plateau is an exposure of pyroclastics that may represent the site of a local vent or fissure eruption.

Mileage	Description
0.0	Junction of Idaho 5 and Idaho 3 in St. Maries.
0.4	Cross the St. Maries River. The quarry on the left is in the Precambrian Wallace Formation, whereas the hills on the right are Striped Peak Formation. Occasional exposures of Miocene basalts occur on lower slopes and in the bottoms of small tributary valleys.
3.5	Miocene basalt exposures identified as the Dodge flow (lower Wanapum Formation) and capped by several flows of the Priest Rapids Member (upper Wanapum Formation). A sequence of Columbia River basalt flows forms a plateau from here to the vicinity of Santa, at the northwest end of the Miocene Clarkia basin.
9.9	The road descends through a sequence of basalts from the Priest Rapids Member of the Wanapum Formation.
10.9	On the right is an excellent example of what appears to be a flow front that moved into relatively clear water. Basalt "pillowing" here shows solid sinuous rods as well as hollow "tubular" structures.

12.5 Cross the St. Maries River, which enters the valley from the east (left). The road follows the valley of John Creek, a tributary of the St. Maries River, which may have served as an outlet of lake water during one or more episodes of lava-damming of the valley to the southeast. Incised meanders of the St. Maries River occur in this area.

13.0 On the left are exposures containing breccias, glass, and vesicular basalts of Miocene age. This may be a vent or fissure site for local eruptions of Miocene volcanic rocks.

14.5 Junction of Idaho 3 and Idaho Highway 6. Turn east (left) toward Santa. Forested hills and mountains east of the St. Maries River valley have been mapped as lower Wallace Formation; those on the west are upper Wallace Formation. The exposure at the highway junction is upper Wallace. From here to Clarkia, the St. Maries River valley seems to follow fault line that parallels other major structural features in the vicinity.

Segment E: Santa to Clarkia Site P-33 (18 miles)

From Santa to Clarkia, the itinerary overlaps the area of the Clarkia basin covered by the *Guidebook and Road Log to the St. Maries River (Clarkia) Area of Northern Idaho* (Smiley and Rember, 1979). The mountains on the east (left) are underlain by lower Wallace metamorphic rocks; those on the west (right) by upper Wallace. The NW-SE trend of the St. Maries River valley between Santa and Clarkia parallels numerous structural features that were established in pre-Clarkia Lake time. Miocene drainage patterns and direction of stream flow (to the northwest) are precisely those of the present day. It is assumed, therefore, that Neogene tectonic activity has been relatively subdued in this part of northern Idaho (Bennett, 1986).

If one can assume a stable geologic setting throughout Neogene time in this area, then the major climatic and biologic changes from Clarkia Lake time (warm-temperate, summer-wet) to the present (cool-temperate, summer-dry) can be attributed to the following factors: (1) a Neogene cooling of global climates; (2) the uplift of the Cascade climatic barrier on the west; (3) the change in Pacific Northwest climate from "maritime" summer-wet to "continental" summer-dry; (4) the expansion of the length and severity of winters and the reduction of a frost-free growing season; and (5) as a consequence of the previous factors, a striking change in Pacific Northwest vegetation, from Miocene domination by eastern American and eastern Asian "exotics" to western American "domestics."

Mileage Description

0.0 Junction, Idaho 3 and Idaho 6. The roadcut exposure is upper Wallace Formation. Proceed east (left) to Santa.

1.0 View eastward to Santa in bottom of the St. Maries River valley. The road descends through Priest Rapids basalt (Wanapum Formation).

1.6 Cross the St. Maries River at Santa.

2.1 Bottomland exposures of Priest Rapids basalt.

3.7 Exposure of Priest Rapids basalt on the left.

6.0 Fernwood. Small roadcuts expose late Neogene stream gravels that rest on Miocene basalt in this area (basalt is in the ravine on the right). The undulatory valley bottom in this northwest (down drainage) end of the basin is the result of such gravel deposition. There is no evidence of Pleistocene glaciation in the St. Maries River valley.

7.5 For the next several miles, the road parallels a small forested ridge in the valley bottom. This is a single basalt flow (Priest Rapids) that may have come from a vent located farther up the valley (up drainage) near Clarkia. Near Clarkia is a small "volcano" of several flows and near-vent pyroclastics formed on top of fossiliferous Clarkia deposits cut by a dike of the same basalt type (Priest Rapids). Valley walls in this area are Belt metasediments (mainly mica schist and quartzite), with lower Wallace on the east and upper Wallace on the west (Griggs, 1973).

10.5 Junction of Idaho 3 and Emerald Creek Road (to the right). This is the mouth of the Emerald Creek embayment of Miocene Clarkia Lake (Figure 2). On the left is a single flow of Miocene basalt resting on fossiliferous Miocene sediments. On the right, the Emerald Creek Road cuts across the basalt ridge mentioned before; basalts occur only near the down drainage end of the Emerald Creek valley. See Smiley and Rember (1979) for details of the iteinary to Clarkia fossil sites P-37 and P-40 (Figure 3). Strip-mining for garnet sand in adjacent flat valley floors temporarily exposes horizontal fossiliferous Clarkia deposits.

12.3 Exposure of lower Wallace rocks on the left, forming the eastern margin of the sedimentary basin.

12.8 **Stop 1: Dodge flow.**

Junction of the St. Maries River and Cedar Creek. On the right is a small State of Idaho Campground. Turn into campground and stop (toilet facilities available). Walk to a small exposure at the left edge of the campsite to see a single flow of basalt and near-vent pyroclastics identified as the Dodge flow of the Wanapum Formation. For more details on Stop 1, see Smiley and Rember (1979) and SMiley (1985).

15.3 Exposure of "flow-front" basalt pillowing (Priest Rapids), as seen earlier in Segment D. Low hills across the valley on the left represent a local pile of several basalt flows with a small quarry exposure of near-vent pyroclastics. Later Neogene erosion has dissected this local basalt pile, forming a ravine that exposes underlying units of fossiliferous Clarkia deposits (Site P-34). The Clarkia beds are intruded by a dike-sill complex of the same basalt type (Priest Rapids), and have been blackened probably by heat.

15.9 St. Joe National Forest work station.

16.3 Road junction of Clarkia on the left. See Smiley and Rember (1979) for details on a side-trip itinary to the "volcano" sites P-34 and P-38 (Figure 3). On the right is a roadcut of upper Wallace metasediments.

18.2 **Stop 2: Clarkia site P-33.**

Fossil collecting site (see Figure 2 and 3). Refer to Smiley and Rember (1979) and Smiley (1985) for details of this locality.

(Retrace itinerary to St. Maries, junction of Idaho 3 and Idaho 5).

Segment F: St. Maries to I-90, Via the Coeur d'Alene River Valley (33 miles)

The itinerary follows along the St. Joe River valley westward for several miles, then turns northward to ascend a metamorphic upland to the valley of the Coeur d'Alene River. Occasional small patches of Miocene basalt are present, mainly as exposures on lower slopes. In the broad valley of the Coeur d'Alene River are many small lakes in various stages of infilling. Between the Coeur d'Alene River valley and I-90 the road traverses a highly fractured Precambrian terrane of Belt metasediments (mainly Prichard).

Mileage	Description
0.0	St. Maries. Junction of Idaho 3 and Idaho 5. Turn north (right) on Idaho 3. The road follows the St. Joe River valley for a distance of about 8 miles. Forested uplands on the north are Precambrian Prichard Formation, with patches of Miocene basalt found largely on lower slopes.
8.4	Exposures of Miocene basalt in roadcuts.
10.0	Exposure of Precambrian Wallace Formation on the left, capped farther up the road by Miocene basalt.
11.0	Junction of Idaho 3 and Idaho Highway 97. Continue north on Idaho 3. Forested hills on the right are Prichard.
15.5	Road begins to descend into the valley of the Coeur d'Alene River.
16.0	Roadcut exposure of Miocene basalt.
17.4	Overview of the Coeur d'Alene River valley, showing a series of lakes in various stages of infilling. Surrounding uplands are fractured Belt metasediments (mainly Prichard).
19.7	Cave Lake Road. Outcrops of Prichard Formation.
26.8	Bridge crossing the Coeur d'Alene River. On the left is a bog of wild rice.
29.7	Rose Lake. The small lake on the right is silting in and is bordered by outward-growing bogs with cattails, sedges, wild rice, water lilies, and various other aquatic plants.
33.1	Junction of Idaho 3 and I-90. The roadcut here is Prichard Formation. Turn west (left) for return to Spokane via Coeur d'Alene.

Segment G: I-90 West to Coeur d'Alene (22.2 miles)

For several miles I-90 follows the Fourth of July Canyon, where numerous roadcuts expose Prichard rocks. As elsewhere, patches of Miocene basalt occur at lower elevations.

Mileage	Description
0.0	Junction of I-90 and Idaho 3.

11.9 Junction of I-90 and Idaho 97. Wolf Lodge Bay at north end of Coeur d'Alene Lake. Idaho 97 is a winding scenic route that follows the east side of Coeur d'Alene Lake southward to a few miles northwest of St. Maries.

15.3 East edge of Coeur d'Alene. Roadcuts here and for the last 3 to 4 miles have exposed Prichard metasediments and Miocene basalts.

22.2 Junction of I-90 and U.S. 95. From here to Spokane, the itinerary retraces Segment A westward down the valley of the Spokane River.

ACKNOWLEDGEMENTS

My thanks go to John Bush and Valerie Chamberlain for reviewing this manuscript.

REFERENCES

Bennett, E. H., 1986, Relationship of the trans-Challis fault system in central Idaho to Eocene and Basin and Range extensions: Geology, v. 4, p. 481-484.

Bickford, M. E., B. P. Rhodes, and D. W. Hyndman, 1985, Age of mylonitization in the southern Priest River complex, northern Idaho and northeastern Washington: Geological Society of America: v. 17, p. 341-342.

Daubenmire, R., 1969, Ecologic plant geography of the Pacific Northwest: Madrono, v. 30, p. 111-128.

Griggs, A. B., 1973, Geologic map of the Spokane quadrangle, Washington, Idaho and Montana: U.S. Geological Survey, Miscellaneous Geological Investigations Map I-768.

Kincer, J. B., 1941, Climate and weather data for the United States, *in* G. Hambridge, editor, Climate and Man: U.S. Department of Agriculture, Yearbook of Agriculture, p. 685-1228.

Miller, F. K., and J. C. Engels, 1975, Distribution and trends of discordant ages of the plutonic rocks of northeastern Washington and northern Idaho: Geological Society of America Bulletin, v. 86, p. 517-528.

Smiley, C. J., editor, 1985, Late Cenozoic History of the Pacific Northwest: Interdisciplinary Studies of the Clarkia Fossil Beds of Northern Idaho: Pacific Division, American Association for the Advancement of Science, California Academy of Sciences, San Francisco, California, 417 p.

———, (in press), Paleolimnology of Miocene Clarkia Lake, Norhtern Idaho, U.S.A., *in* Jane Gray, editor, Paleolimnology: Elsevier Science Publishers, Amsterdam, The Netherlands.

Smiley, C. J., and W. C. Rember, 1979, Guidebook and road log to the St. Maries River (Clarkia) fossil area of northern Idaho: Idaho Bureau of Mines and Geology Information Circular 33, 27 p.

———, 1985a, Physical setting of the Miocene Clarkia fossil beds, northern Idaho, *in* C. J. Smiley, editor, Late Cenozoic History of the Pacific Northwest: Interdisciplinary Studies of the Clarkia Fossil Beds of Northern Idaho: Pacific Division, American Association for the Advancement of Science, California Academy of Sciences, San Francisco, California, p. 95-112.

———, 1985b, Composition of the Miocene Clarkia flora, *in* C. J. Smiley, editor, Late Cenozoic History of the Pacific Northwest: Interdisciplinary Studies of the Clarkia Fossil Beds of Northern Idaho: Pacific Division, American Association for the Advancement of Science, California Academy of Sciences, San Francisco, California, p. 11-31.

Smith, G. R., and R. L. Elder, 1985, Environmental interpretation of burial and preservation of Clarkia fishes, *in* C. J. Smiley, editor, Late Cenozoic History of the Pacific Northwest: Interdisciplinary Studies of the Clarkia Fossil Beds of Northern Idaho: Pacific Division, American Association for the Advancement of Science, California Academy of Sciences, San Francisco, California, p. 85-94.

Smith, G. R., and R. R. Miller, 1985, Taxonomy of fishes from Miocene Clarkia Lake beds, Idaho, *in* C. J. Smiley, editor, Late Cenozoic History of the Pacific Northwest: Interdisciplinary Studies of the Clarkia Fossil Beds of Northern Idaho: Pacific Division, American Association for the Advancement of Science, California Academy of Sciences, San Francisco, California, p. 75-84.

Chapter Two
Idaho Batholith and Accreted Terranes

Yellowjacket Lake lies in glaciated terrain of the Idaho batholith. *Photograph credit courtesy of Idaho Division of Travel Promotion.*

Formation of the Northern Idaho Batholith and the Related Mylonite of the Western Idaho Suture Zone

Donald W. Hyndman[1]

INTRODUCTION

This field trip guide outlines the Cretaceous evolution of the northern Idaho batholith and its surrounding high-grade regional metamorphic rocks at the westernmost margin of Precambrian continental North America. The present exposures are at midcrustal levels, perhaps transitional in characteristics to those exposed at deeper levels in some Archean basement terranes.

On the west, the Wallowa-Seven Devils terrane of volcanic arc and oceanic affinity juxtaposes the truncated margin of western North America across the western Idaho suture zone, which is virtually coincident with the strontium isotope 0.704 initial ratio boundary.

[1] *Department of Geology, University of Montana, Missoula, Montana 59812.*

EVOLUTION OF THE IDAHO BATHOLITH ENVIRONMENT

The Idaho batholith is emplaced into the Precambrian continental crust of North America, just east of a major collisional suture zone bounding the Wallowa-Seven Devils terrane or microcontinent to the southwest (Figure 1). Beginning about 120 Ma, the Wallowa-Seven Devils terrane arrived in western Idaho, colliding with the Precambrian rifted margin of old North America, and completing the docking by the intrusion of quartz dioritic plutons 85 to 82 Ma (Lund and others, 1985; Snee and others, 1987). The boundary is marked by a major mylonitic suture zone, called the western Idaho suture zone. This zone trends northward in western Idaho near the western edge of the Idaho batholith, then curves westward near the town of Orofino at latitude 46°30'N,

Hyndman, D. W., 1989, Formation of the northern Idaho batholith and the related mylonite of the western Idaho suture zone, *in* V. E. Chamberlain, R. M. Breckenridge, and Bill Bonnichsen, editors, Guidebook to the Geology of Northern and Western Idaho and Surrounding Area: Idaho Geological Survey Bulletin 28, p. 51-64.

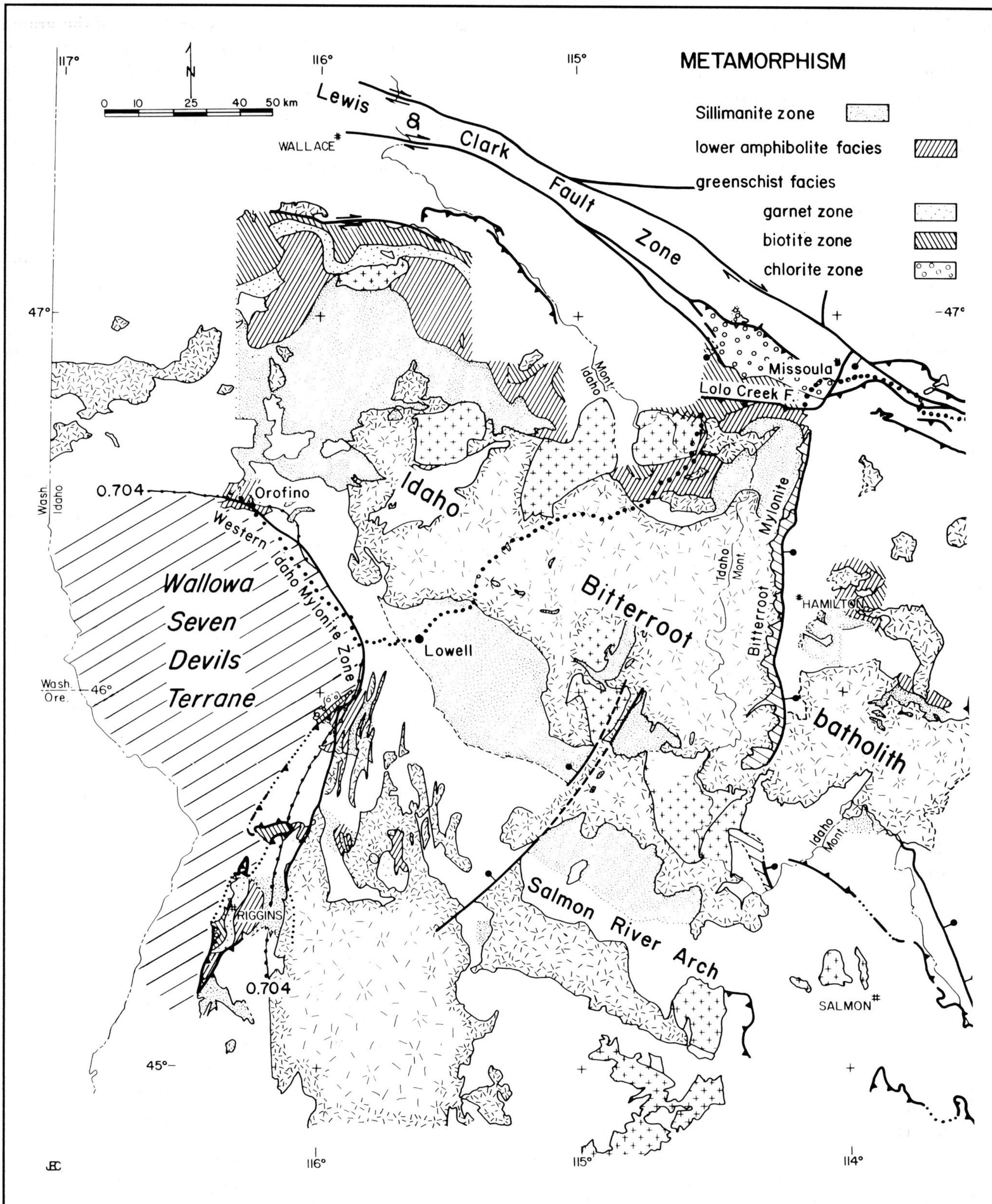

Figure 1. Tectonic and regional metamorphic map of the Idaho batholith region. Route of the field trip is shown as a dotted line (modified from Hyndman and others, 1988a)

and disappears beneath the Miocene Columbia River basalts. At the bend, the microcontinent was being thrust northeastward beneath North America. The present plunge of the mylonitic lineation at the bend is now 50 degrees northeastward. At least 85 kilometers of northwest over southeast transport are suggested by the deformation of dikes in the mylonite zone (Strayer, 1988). The suture zone closely follows the boundary between generally mafic rocks of oceanic affinity on the west and generally felsic rocks of continental affinity on the east and between $^{87}Sr/^{86}Sr$ initial ratios of less than 0.704 on the west and greater than 0.706 on the east (Armstrong and others, 1977; Fleck and Criss, 1985).

Semipelitic to quartzose and impure calcareous sedimentary rocks of the Proterozoic Belt Supergroup, underlying most of northwestern Montana and northern Idaho, were subjected to regional burial metamorphism to the greenschist facies (Norwick, 1972). Metamorphism reached lower grades higher in the stratigraphic section. During Mesozoic time, rocks adjacent to the northern Idaho batholith were deformed and subjected to regional-dynamothermal metamorphism to grades reaching as high as the sillimanite zones of the amphibolite facies (cf. Hyndman and others, 1988a), an effect that extends as much as 70 kilometers from the batholith contact (Figure 1). In contrast to the massive, spotted fabric of the Precambrian burial metamorphism, the Mesozoic regional metamorphism formed well-foliated schists and gneisses. Metamorphic grades increase progressively, in most areas, towards the batholith. Locally, however, the isograds are cut at low angles by the batholith contact, demonstrating that the latest effects of intrusion postdated regional metamorphism. Mineral assemblages that include quartz, muscovite, biotite, staurolite, sillimanite, and kyanite suggest that rocks north and northwest of the batholith were formed at depths greater than 25 kilometers, but somewhat less to the southeast (Rice, 1987; Hyndman and others, 1988a). Melting formed anatectic migmatites at the highest metamorphic grades, suggesting that this regional metamorphism accompanied crustal melting that formed the Idaho batholith (Hyndman, 1981).

The main-phase units of the northern Idaho batholith, or Idaho-Bitterroot batholith (Figure 2), are granite to granodiorite emplaced in latest Cretaceous time between about 65 and 80 or possibly 85 Ma. The timing, immediately after collision along the western Idaho suture zone, and the proximity of the batholith to that zone suggest that the subduction responsible for creating the batholith occurred west or southwest of the western Idaho suture zone. The 14,000-square-kilometer Idaho-Bitterroot batholith was apparently emplaced into a broad synclinorium trending southeastward in the Proterozoic Belt Supergroup. The batholith rocks are felsic, medium-grained, and massive or nearly so. Much of the granite along the Lochsa River, the deep line of section followed by the field trip in this part of the batholith, is relatively homogeneous mineralogically and chemically (Hyndman, 1984); elsewhere, however, the batholith shows considerable variation within the range of granite and granodiorite (Hyndman, 1983; Toth, 1987; Reid, 1987).

The western border zone, about 20 percent of the width of the Idaho batholith, consists of somewhat earlier intrusions of biotite-hornblende quartz diorite to tonalite. The rocks are massive to generally foliated on steep surfaces trending approximately parallel to the batholith contact. Sheets and other bodies of tonalitic and quartz diorite orthogneiss occur elsewhere in the border and deeper zones of the batholith (cf. Taubeneck, 1971; Chase, 1973; Reid and others, 1979; Wiswall and Hyndman, 1986; Hyndman and Foster, 1988). These relatively mafic rocks may mark the deepest levels of the Idaho batholith, as also inferred for the southern end of the Sierra Nevada batholith (Ross, 1985).

Extensive and voluminous swarms of synplutonic basaltic andesite to andesite dikes and small areas of gabbroic complex cut the batholith (Hyndman and Foster, 1988). In one 10-kilometer-wide section along the Lochsa River valley, we have measured 20 percent dikes, by volume. Mafic synplutonic dikes are also widespread in the tonalite/quartz diorite of the western border zone of the batholith.

Intrusive relations indicate that the mafic dikes in the Idaho-Bitterroot batholith were emplaced in the same period as the main-phase granites. Evidence for such a synplutonic relationship includes the following observations (Hyndman and Foster, 1988; Foster and Hyndman, in press):

1. Mafic dikes cut the host granite as tabular, sharply bounded bodies; some are stretched, thinned, segmented, or dismembered or are folded in undeformed granite.
2. Many of the mafic dikes show foliation, lineation, and mylonitic texture (see Figure 3) imposed by postemplacement movement of the granite. Some are foliated obliquely to the walls of the dike.
3. Mafic dikes grade into rounded blobs or mafic schlieren in the granite (see Figure 4).
4. Mafic dikes are cut by tabular or ptygmatic dikes of the host granite, are metamorphosed by heat from the host granite, or contain late alkali feldspar megacrysts like those in the granite.
5. Granite within a meter or so of some of the dikes is contaminated and more mafic than normal Idaho batholith granite.

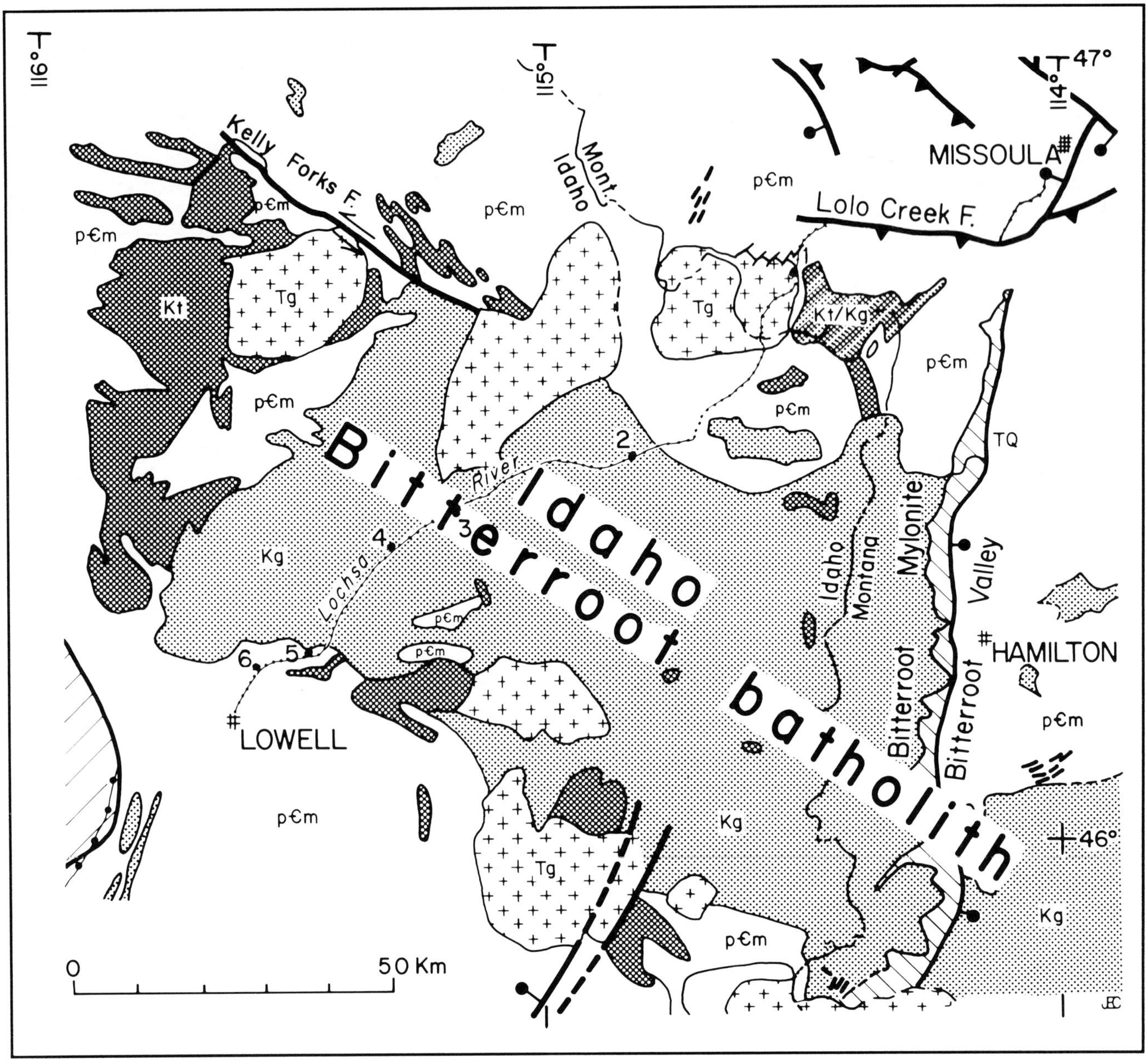

Figure 2. Geologic map of the northern Idaho batholith. p€m = metamorphosed Precambrian metasedimentary rocks; Kt = Cretaceous tonalite/quartz diorite; Kg = Cretaceous granodiorite/granite main-phase units of the batholith; Tg = Eocene plutons emplaced at shallow depths; TQ = Tertiary-Quaternary valley-fill sediments (after Hyndman and Foster, 1988). Field trip stops are numbered.

6. Composite dikes show small-scale intermingling between granite and andesite or rounded blobs of basalt with thin leucogranite rims.

7. The complete range of chemical composition of the synplutonic dikes, from basaltic andesite to dacite, lies along a linear, major-element, chemical-mixing trend which includes the composition of the enclosing granite to granodiorite. Magma mingling and mixing are supported by megascopic to abundant microscopic disequilibrium textures in intermediate rock types.

Given the nature of their interaction with the enclosing granite, synplutonic mafic magmas appear to have been injected at early, intermediate, and late stages during crystallization of the granite. Since voluminous, high-temperature, mafic magmas apparently were emplaced throughout the crystallization time of the magmas of the Idaho-Bitterroot batholith, they must have added consid-

Figure 3. Mylonitic mafic dike enclosed in relatively undeformed granite. Dike was deformed in still molten granite. U.S. 12, mile 124.9, below road.

Figure 4. Fine-grained mafic dike inclusions in intermediate composition and grain size zone, enclosed in medium-grained granite. U.S. 12 mile 124.75, below road.

erable heat. As suggested by Hyndman and Foster (1988), the mafic magmas, rising above the subduction zone, probably provided the additional heat required to melt the lower continental crust to form the granitoid magmas of the batholith.

Pegmatites and minor aplites occur in several environments. A few early, concordant muscovite-biotite-quartz-feldspar pegmatites pinch and swell and show 1 to 3 millimeter-sized biotite selvages in the pelitic to quartzofeldspathic country rocks. These probably formed by anatexis during regional metamorphism. Most of the pegmatitic dikes were emplaced by injection (Chase and Johnson, 1977) during or after regional metamorphism; many are probably associated with emplacement of the main-phase units of the batholith.

Many tabular, postmetamorphic muscovite-quartz-feldspar pegmatites, 5 to 25 centimeters thick, have injected steeply dipping planar fractures. Another group of tabular pegmatites or aplites, 1 centimeter to 2 meters thick, dips 5 to 25° W, NW, or N, regardless of the dip or degree of deformation of the enclosing rock. This second group of subhorizontal pegmatitic sheets is emplaced only into granodiorites and granites of the western 16 kilometers of the Lochsa River valley section (west of milepost 120) and extends for only 1.3 kilometers into the country rocks (Hyndman, 1983). The surfaces of the pegmatite dikes are marked by slickenside striae plunging down dip to the northwest, indicating postemplacement slip and probably injection along the shear-surfaces during emplacement.

Several stages of deformation are recognized in the Idaho batholith environment (cf. Hietanen, 1961; Chase, 1973; Reid and others, 1973; Nold, 1974; Reid and others, 1979; Childs, 1982; Wiswall and Hyndman, 1986; Rice, 1987; Hyndman and others, 1988a). Tight isoclinal folds with strong axial-plane schistosity were formed in deformation D1, but they are difficult to find. A few large folds verging away from the batholith have been described. The major amphibolite-facies regional metamorphism surrounding the batholith, and described above, produced the axial-plane schistosity of D1 folds. Open to nearly isoclinal folds of deformation D2 refold the dominant schistosity and develop a new weak to locally strong axial-plane schistosity of muscovite and biotite. These folds tend to have nearly vertical plunges near the batholith contacts, perhaps as a result of rotation during rise of the batholith magmas during D3. Upright and relatively open concentric and flexural slip folds of the earlier metamorphic schistosity formed during D3 deformation, probably during the final stages of forceful emplacement of the batholith. No new axial plane schistosity formed. The rapid rise of the batholith may have accompanied emplacement.

An overall model for the formation of the northern Idaho batholith is as follows. The Wallowa-Seven Devils terrane of eastern Oregon and western Idaho docked in late Cretaceous time near the present western edge of the Idaho batholith, a boundary marked by mylonite of the western Idaho suture zone and the Sr-isotope initial-ratio boundary. Subduction on a separate zone southwest of the terrane produced the mafic western border zone between about 120 to 80 million years ago and generated the felsic main-phase magmas of the Idaho batholith about 80 to 65 million years ago. Partial melting formed basaltic andesite magmas in the mantle wedge above the subducted oceanic plate, and those high-temperature magmas rose into the overlying crust. Outboard of old continental North America, they crystallized to form tonalite and quartz diorite of the Kamiah plutonic complex, a large area of igneous rocks in the northern part of the Wallowa-Seven Devils terrane (Table 1). Where the mafic magmas rose into the old granitoid continental crust, they heated the crust to cause regional metamorphism and ultimately to cause secondary melting to form the granodiorite and granite magmas of the Idaho batholith. The synplutonic mafic dikes now preserved in the Kamiah plutonic complex and the Idaho batholith formed by the continued rise of mafic magmas, probably from the same subduction zone.

Movement represented by the mylonites of the western Idaho suture zone continued during the late stages of crystallization of the Kamiah plutonic complex. Increase in rock strength accompaning the final crystallization of the complex caused the end of mylonitic movement along the suture (Hyndman and others, 1988b).

ROAD LOG

The Idaho batholith field trip traverses a well-exposed cross-section of the northern Idaho batholith, briefly examines the broad aspects of this deep-seated granitoid batholith and its regional metamorphic country rocks, and considers the role of the synplutonic mafic magmas from the mantle in providing heat for melting the continental crustal rocks to form the more felsic main-phase units of the batholith. The route passes the following features:

1. The low- to high-grade regional metamorphic rocks of the eastern border zone.
2. The main phase granodiorite and granite of the batholith.
3. Fine-grained synplutonic dikes of basaltic andesite to andesite which cut the batholith and make up about 20 percent of the batholith's volume; their complex mixing relationships with the batholith magmas.
4. The high- to medium-grade regional metamorphic rocks and sheetlike intrusions of the western border zone.
5. Injection migmatites, the border zone rocks.
6. Some of the structures which relate to emplacement of the batholith.
7. Early western border zone tonalites.

Day 1: The Idaho Batholith, Mafic Synplutonic Dikes, and Country Rocks

Stops are keyed to highway mileposts. The trip begins in Missoula, Montana, and travels to Lowell, Kooskia, and Orofino, Idaho, via U.S. Highway 12 (U.S. 12).

Milepost	Description
90.1	Begin trip on U.S. Highway 93 at bridge across Bitterroot River at south edge of Missoula.
88.4 to 87.4	Roadcuts in red to pale brownish to pale greenish, well-layered mudstones of Proterozoic Missoula Group of the Belt Supergroup. These rocks show ripple marks and mudcracks indicative of deposition in shallow water, either a tidal flat or river floodplain.
83.4 to 32.6	Junction U.S. 93 and U.S. 12 at Lolo. U.S. 93 joins U.S. 12 at milepost 32.6. Travel west on U.S. 12. U.S. 12 follows the Lolo Creek fault, a steep reverse fault with the south side up. Wallace Formation (underlies Missoula Group) sedimentary rocks of the biotite zone of the greenschist facies are on the north or downthrown side; Ravalli Group quartzite (underlies Wallace Formation) in the amphibolite facies is on the upthrown side. Prichard Formation semipelitic meta-sedimentary rocks of the lower Belt Supergroup, underlying the Ravalli Group farther south, are in the sillimanite zone bordering the northern Idaho batholith. Continue along U.S. 12, for 25 miles west of Lolo, to Lolo Hot Springs at milepost 7.5.
26.0 to 25.1	Mt. Lolo (elev. 9075 feet), visible on the skyline to the south (left), is in sillimanite zone Prichard Formation metasediments (see Wehrenberg, 1972). The northeast corner of the Idaho batholith is 10 miles farther south of Mt. Lolo.
22.0	Pale gray to white, poorly bedded quartzite of the Ravalli Group.
18.1 to 8.1	Roadcuts are mostly in thinly layered Wallace Formation dolomitic siltstones with variable degrees of rust-colored weathering.
7.5	**Stop 1: Lolo batholith at Lolo Hot Springs.** Large rounded exposure of the Eocene Lolo batholith, just inside the northeast contact of the batholith. It is a massive, coarse-grained, shallow, hypersolvus, A-type granite with miarolitic cavities. The rock is dominated by coarse, pink, perthitic alkali feldspar with lesser gray quartz. Terminated grains of quartz and alkali feldspar line the cavities. A few minute grains of purple fluorite are also present. This body, emplaced at very shallow levels, erupted rhyolitic volcanic rocks in a caldera environment to the west (Simpson, 1985).

0.00/ 174.35 Lolo Pass on the Montana-Idaho border. Montana milepost 0.0 equals Idaho milepost 174.35. Idaho mileposts follow.

170.8 Small metamorphosed gabbroic layered intrusion studied by Jens (1974). Black ultramafic layers of diopside and hornblende alternate with gray mafic layers of foliated amphibolite. The intrusion is faulted and cut by many discordant pegmatites, but its age, though pre-metamorphic, is uncertain.

169.75 Just northeast of bridge across Crooked Fork Creek.

Exposure of quartz diorite or granodiorite of the Brushy Fork stock (Nold, 1974) cutting Ravalli Group metaquartzite. It is a medium-grained biotite quartz diorite or tonalite with about 20 percent biotite. The foliation is about S. 67° E. / 71° NE. The granitoid rock is cut by a discordant muscovite granite pegmatite/aplite dike.

A 1-meter-long xenolith of muscovite-biotite quartzite has foliation and layering concordant to the contact and to the foliation in the granitoid intrusion. Crosswarps of the schistosity are at about 90 degrees.

165.8 Calc-silicate paragneiss, probably Wallace Formation, has been regionally metamorphosed to the amphibolite facies. Layering at N. 74° E. / 45° N. is followed by a concordant muscovite-biotite-quartzofeldspathic alaskite dike about 15 centimeters thick and cut by a discordant dike 45 centimeters thick. A vertical granite dike, 3 or 4 meters thick, has reacted with diopside to form black hornblende. Several fine-grained mafic dikes about 60 centimeters to 1 meter thick dip gently northeast. These are internally foliated and contain small biotite porphyroblasts.

164.05 A rock slide made up of large blocks of well-layered metamorphosed Wallace Formation with actinolite-diopside-plagioclase layers and biotite quartzite layers. The largest block above the road shows a well-developed dilational pegmatite with thin aplite and pegmatite borders. Smaller blocks just below the road and slightly east show a rippled-looking surface. Are these preserved ripple marks or the intersection of spaced cleavage with the layering and schistosity?

159.5 0.1 mile northeast of Papoose Creek.

Nearly vertical, dark gray, synplutonic, mafic dikes cut granite and metasedimentary rocks of the Idaho batholith border zone. The dikes trend nearly parallel to the roadcut. The dikes are metamorphosed and now consist of biotite-plagioclase rock. Strong foliation and slight lamination in the dikes are essentially parallel to the dike walls. Granite and pegmatite dikes up to 1.2 meters thick cut the mafic dikes.

156.45 **Stop 2: Complexely deformed early phase of the batholith.**

Small pullout just east of curve on south side of road.

Megacryst-rich biotite granodiorite, apparently an early, complexly deformed phase of the Idaho batholith, shows complex veining by granitic dikes; the most photogenic exposures are in blocks below the road and slightly east of the highway pullout. K-feldspar megacrysts average about 2.5 by 4 centimeters and have white plagioclase rims. They are subhedral and make up 5 to 10 percent of the rock; locally 15- to 20-centimeter-wide zones contain up to 35 percent megacrysts. The groundmass contains no K-feldspar. Consider alternative origins for the K-feldspar megacrysts.

(a) Early-formed phenocrysts? Note that the magma/rock composition on an An-Ab-Or ternary does not plot in K-feldspar-first field. Why no K-feldspar in the groundmass (no apparent peritectic relationship to eliminate K-feldspar)?

(b) Post-magmatic porphyroblasts? Note that one or two straddle granodiorite/country rock contacts. If isochemical metamorphic porphyroblasts, why is only the K-feldspar so coarse? Why the concentric inclusions? If metasomatic porphyroblasts, why are K-feldspar porphyroblasts essentially confined to the granodiorite and not in the country rock?

(c) Late-magmatic grains (crystallization order plausible from experimental T-X_{H2O} diagrams of Whitney, 1975), large because of higher water content of magma in late stages? Consider why they are subhedral and some have concentrically enclosed inclusions of other minerals in the rock. Have they replaced preexisting groundmass?

I prefer explanation (c). The buildup of water in the late stages of crystallization would tend to inhibit nucleation of K-feldspar which crystallizes late in this composition of magma. It would foster rapid migration of constituents to the few growing grains, as in pegmatites. In addition, the activity of water could aid in the replacement of preexisting groundmass grains. In this explanation, the K-feldspar megacrysts are neither typical phenocrysts (which would crystallize early from the magma) nor porphyroblasts (which crystallize in the solid rocks by replacement of preexisting grains).

Foliation at S. 40° E. / 74° SW. is marked by K-feldspar megacrysts, plagioclase, and biotite. Irregular criss-crossing veins of biotite-quartz-feldspar pegmatite and alaskite are 1 to 10 centimeters thick. Dikes of layered medium-grained granite are up to 1 meter thick and locally up to 3 and even 6 meters thick. The subtly graded layers in the dikes have a variable mafic content and grain size, with grains decreasing in size upwards. The settling of mafic grains in a viscous granitoid magma may have been fostered by shear in the magma.

148.9 Granite continues from milepost 156.5 to 144.55. A big roadcut in dark gray, biotite-rich quartz diorite orthogneiss is laced by many criss-crossing pegmatite and alaskite dikes 1 to 60 centimeters wide. Plagioclase making up about 40 percent of the rock occurs as subhedral, white, 4 to 7 millimeter long grains. Scattered translucent megacrysts of K-feldspar have whiter rims of plagioclase 1 or 2 millimeters thick. Quartz diorite orthogneiss cuts layered paragneiss (Belt rocks?). One planar, subhorizontal fine-grained mafic dike, 30 to 60 centimeters thick, cuts the other rocks.

141.9 Main-phase biotite granite with 1-2 percent K-feldspar megacrysts. A few faint schlieren are at N. 55° W. / 37° NE. Very fine-grained, dark gray mafic dikes, 15 to 30 centimeters thick, cut the granite. The mafic dike at the southwest end of the outcrop (near milepost 141.85) shows foliation oblique to its borders.

141.85 Essentially massive megacryst-bearing, main-phase granite. Faint 2- to 5-centimeter-thick schlieren are at N. 40° W. / 45° NE. Very irregular 15- to 60-centimeter-wide mafic dikes cut the granite on curved fractures. One mafic dike contains a 2 by 5 centimeter xenolith of medium-grained granite.

139.2 Granodiorite complex exposed to milepost 139.1 shows complex interactions between granodiorite and fine-grained mafic magmas.

Medium-grained biotite granodiorite is inhomogeneous and has more felsic zones, more mafic zones, and granitic pegmatites. The more mafic parts of the granodiorite show swirled, foliated textures and granitic patches and are cut by fine-grained, light gray felsic dikes and thin dikelets. The granodiorite is in curving, irregular contact with a finer grained, more mafic phase that is cut by pod-shaped masses and dikes of granite. The granodiorite is cut by mafic dikes and contains mafic inclusions. The mafic rocks are, in turn, cut by pegmatites.

Both granite and granodiorite are cut by planar joints, about 80 degrees to the foliation, that show pink medium-grained K-feldspar alteration zones about 2 centimeters wide. This alteration was probably associated with Tertiary intrusions nearby.

134.3 **Stop 3: Synplutonic mafic dikes.**

Stop at pullout on south side of road.

A sample of quartz monzodiorite at milepost 134.15 has provided a U-Pb zircon lower intercept of 66 million years; this is presumed to be the time of crystallization (Schuster and Bickford, 1985).

The synplutonic mafic dikes in this exposure show a broad range of synplutonic interrelationships, including mutual intrusion, mingling, and mixing of magmas. Three main dikes are exposed in the roadcut. The dike to the southwest is thin, poorly exposed, and mylonitic. The thick (3.2 m) dike in the center of the cut is fine-grained, gray basaltic andesite to andesite with some areas containing 1- to 2- millimeter-diameter mafic spots. The dike is very inhomogeneous and consists of many rounded inclusions or blobs of andesite with thin rims of felsic material and a matrix of andesite. It contains one rounded inclusion of granite, 5 centimeters in diameter.

The mafic dike on the northeast curves and steepens upward into a thick sheet or sill, which shows clear synplutonic relationships with the host granite. It is somewhat finer grained and chilled against the intruded granite. Numerous granitic dikes and ptygmatic granitic pods injected into the thick mafic sheet are contorted or pinch and swell. Some seem to surround pillows of andesite. Below the mafic sheet, the granitic rock is contaminated to a very inhomogeneous granodiorite to quartz diorite composition with many felsic and mafic inclusions. The more mafic granitoid rocks from high on the outcrop can be sampled in the loose material in the borrow pit. A loose block half-way down to the river and 5 or 6 meters downstream from the thickest dike shows fine-grained dark gray dike rock laced with medium gray, more-felsic dikelets. Several other blocks nearby show the same.

128.55 to 128.7 Synplutonic gabbro/metaperidotite complex. The brownish weathering peridotite or hornblende pyroxenite in the central part of the long exposure may be a differentiated magma chamber of the mafic dike magma(?). Hornblende-rich (as much as 50 percent hornblende) metagabbro is also present. To the southwest is a hornblende diorite to biotite granodiorite injection zone. The mafic mineral content is highly variable; the lighter-colored dioritic to granodioritic phases contain darker, sharply bounded inclusions of hornblende-rich diorite. It probably was a zone of mixing between mafic and felsic magmas. Farther southwest these rocks are cut by granodiorite to granite.

128.45 State of Idaho highway maintenance shop.

125.0 **Stop 4: Synplutonic dikes, magma mingling and magma mixing.**

Stop at pullout on south side of road.

This exposure displays numerous synplutonic mafic dikes, making up about 20 percent of the outcrop, and a broad range of magma mingling and mixing relationships in inhomogeneous granodiorite. Walk through milepost 125.0-124.9 and consider the total percentage of mafic dike material. The mafic dikes show sharp to moderately sharp, but irregular to planar, contacts to the granitoid rock; some dikes are foliated and veined at a large angle to contacts; 3-millimeter-long mafic phenocrysts are now altered. The granodiorite is inhomogeneous. Some dikes are lineated and have streaks of granitoid material and numerous small mafic spots—pseudomorphs of small mafic phenocrysts. Some dikes show an intermingling of andesite and granite. Granitoid areas in some dikes are more mafic, having been contaminated by the mafic dike magma. One dacite dike shows veins, streaks, spots, megacrysts, inclusions of felsic material, and a xenolith of a mafic dike. One area of small dikes grades into schlieren in the granodiorite. Angular xenoliths of meta-andesite in the granodiorite contain mafic spots 1-2 millimeters in diameter. Note that boulders near the river at mile 124.95 show mafic dikes that grade into pillow-like masses and schlieren.

About 400 meters downstream from milepost 125 (23 meters downstream from Noseeum Creek which flows through a huge pipe under the road), and below the road, is a block of medium-grained granite with fine-grained, dark gray, elongate inclusions of mafic dike material, which are surrounded by a "mixed" zone of intermediate composition and grain size.

122.4 The dike rock is near the big pine tree on outside bend of road. Pull off on south side of road at 122.5. This large area of mafic dike rock is marginally exposed, but shows much interaction with the granite. Eight meters upslope are good exposures of comingled mafic and felsic magmas. The granite contains a few subrounded mafic xenoliths, 2 to 10 centimeters in diameter.

118.6 Mafic dikes with xenocrysts(?). Massive dikes of intermediate composition have very small gabbroic xenoliths, now altered white. One dike has chilled margins. The mafic dikes cut biotite granite with K-feldspar megacrysts. Country rock to the west, across a steeply dipping fault, is a foliated migmatitic calc-silicate gneiss with abundant diopside. Alaskite dikes 1 to 8 centimeters thick criss-cross the outcrop of gneiss; their most common dip is 15° NW.

116.0 Massive main-phase granite. A large, irregular xenolith of contorted biotite-rich gneiss is enclosed in the granite which also shows schlieren. A few gently dipping pegmatite dikes, 2 to 45 centimeters thick, cut the granite and are discordant to the mafic schlieren. A steeply dipping, dark gray, fine-grained andesite dike, 8 to 12 centimeters thick, cuts the granite. A mafic dike appears as a large angular patch, with a chilled border on one side and a sheared margin on the other side. A pod of fine-grained rhyolite (Eocene?) about 2 meters thick dips 10°-20°NE. and has a wavy contact at the granite.

112.5 Biotite quartzofeldspathic gneiss and veined gneiss with foliation at S. 73° E. / 78° NE. and main-phase granite, both cut by a subhorizontal layered granite sheet about 1 to 3 meters thick. The layered granite sheet shows grading with mafic grains concentrated towards the base of each 2- to 30-centimeter-thick layer. The grading may result from movement on a subhorizontal shear in the late stages of crystallization of the granite (a late-magmatic "mylonite" zone). The layered granite sheet is cut by a gently dipping muscovite-quartz-feldspar pegmatite that is, in turn, cut by a fine-grained greenish white rhyolitic dike.

111.3 Main-phase granite of the Idaho batholith. It contains about 10 percent megacrysts of K-feldspar that typically are about 2 or 3 centimeters long and 1 centimeter wide. The rock contains some gneissic schlieren with steeply plunging folds.

110.5 Intrusion breccia at contact of granodiorite and biotite-rich country rock with contorted, near-vertical foliation. The folds plunge on very steep axes. Several pegmatite dikes, 5 to 75 centimeters thick, dip gently northwest, and a gently southwest-dipping pegmatite dike about 2 meters thick cuts the intrusion breccia. Just to the east is massive fine- to medium-grained granodiorite containing many wavy, streaky xenoliths and schlieren.

109.3 to 109.25 **Stop 5: Calc-silicate gneiss.**

Pull off next to river at milepost 109.4.

Calc-silicate gneiss, probably equivalent to the Wallace Formation, is steeply foliated. It is cut by subconcordant, irregular veins of streaky hornblende granodiorite and a few 8-centimeter-thick aplite veins. The granodiorite and diopside-rich gneiss have reacted to form hornblende zones. In some places the diopside-rich gneiss is tightly folded on vertical axial planes, with fold axes plunging about 50 degrees east-southeast. The veining postdates the folding.

105.3 The diopside-bearing calc-silicate gneiss continues southwest to here.

104.75 **Stop 6: Early megacryst-rich tonalitic unit of the batholith.**

This stop exhibits several intrusive relationships that can be placed in sequence: (1) the intrusion of the metasediments by the granodiorite and mafic dikes, (2) the crenulation of the granodiorite and foliation of the mafic dikes, and (3) the intrusion of later granitic dikes. The best exposures are in blocks between the road and the river. This megacryst-rich biotite granodiorite is a tonalitic rock with K-feldspar megacrysts. K-feldspar megacrysts also exist in some granitic dikes that cut the granodiorite. This unit cuts and locally contains biotite-quartz schist xenoliths that may be metamorphosed Belt rocks. Megacrysts and biotite define a steep foliation; small folds in the schist plunge 63° to S. 55°E. The foliation is crenulated on a decimeter scale and becomes strongly sheared and even mylonitic on some limbs; the megacrysts have become augen in the sheared areas. Undeformed discordant granitic dikes up to 40 centimeters thick cut the granodiorite parallel to the axial planes of the foliation crenulations. A few mafic dikes show mutually cross-cutting relationships with granodiorite, and they approximately parallel the most attenuated, or sheared, limbs of crenulations in the granodiorite. The mafic dikes are themselves foliated.

The sequence of events appears to be the intrusion and foliation of the granodiorite, overlapping with an intrusion of mafic dikes, and probably overlapping with the crenulation of the granodiorite. These events were followed by the intrusion of granitic dikes parallel to the crenulations. If the biotite-quartz schist is metamorphosed Belt rock, the granodiorite cannot be pre-Belt basement; therefore, it presumably is early-phase Idaho batholith.

Tonalitic augen gneiss occurs farther southwest to milepost 104.5; then biotite-quartz schist and gneiss occur, and then some biotite-rich tonalitic gniess continues to the southwest.

103.4 Felsic Eocene dikes containing euhedral phenocrysts of quartz.

103.15 Foliated biotite-tonalitic orthogneiss shows mafic schlieren, xenoliths, and diffusely bounded pegmatitic phases.

102.1 Migmatite. A medium-grained, laminated, biotite-quartz-feldspar gneiss with foliation at S. 64° E. / 76° NE. is cut by a concordant biotite-quartz-feldspar pegmatite, 2 meters or more thick, and by a steeply dipping, dark gray, fine-grained, 60-centimeter-wide andesite dike with 30 percent plagioclase phenocrysts.

Southwest of milepost 100.8 the rock is mostly biotite tonalitic gneiss.

96.85 Near bridge across Lochsa River at Lowell.

A medium-grained kyanite-biotite-quartzofeldspathic gneiss is foliated at S. 35° E. / 52° SW. The schistosity is crenulated on the scale of the 1- to 3-millimeter-diameter grains.

Metasedimentary and tonalitic rocks in this region west of the contact of the Idaho-Bitterroot batholith have been described by Greenwood and Morrison (1973) and Reid and others (1979).

Day 2: Kamiah Complex Quartz Diorite/Tonalite and Mylonite of the Western Idaho Suture Zone

At Lowell, the Lochsa River joins the Selway River to form the Clearwater River which U.S. 12 follows downstream.

Farther west, the Miocene flood basalts of the Columbia River Plateau cap the more subdued topography above the river. They filled the ancestral valley of the Clearwater River and are in places exposed near the road. The following section is modified from Strayer (1988).

Milepost Description

67.6 U.S. 12 crosses the Clearwater River and enters Kamiah.

The late Cretaceous Kamiah plutonic complex lies west of the $^{87}Sr/^{86}Sr$ 0.704 line (Figure 1), which marks the boundary between old continental North America and the more mafic Mesozoic accreted terranes of volcanic arc and oceanic affinity to the west. Given available data, the field trip appears to cross the 0.704 line near Kamiah. Structurally the boundary is marked by mylonites of the western Idaho suture zone. The Kamiah plutonic complex is dominated by quartz diorite and tonalite cut by fine-grained mafic, synplutonic basaltic dikes. The dikes typically have steep dips but no regular trend (Hietanen, 1962). Mutually crosscutting relationships between the basaltic dikes and the enclosing quartz diorite/tonalite resemble those examined in the Idaho-Bitterroot batholith on the previous day of this trip. The Kamiah plutonic complex is well exposed along U.S. 12 from Kamiah to west of Orofino.

64.9 **Stop 7: Kamiah plutonic complex.**

Pull off at historical marker for Long Camp and Asa Smith Mission and walk 0.1 mile west.

This stop exhibits the synplutonic nature of the basaltic dikes which cut the quartz diorite/tonalite of the Kamiah plutonic complex. These dikes are used as markers to indicate the degree of strain in the mylonite of the western Idaho suture zone at stops 8 and 9. Here at Stop 7, fine-grained, dark gray, mafic dikes cut the medium-grained, pale gray, quartz diorites and tonalites of the Kamiah complex. Boundaries of the dikes tend to be sharp and in some places show chilled borders against the intruded tonalitic rocks. Some dikes have planar, parallel contacts; others are somewhat cuspate or irregular. Dilational stringers and dikes from the tonalitic rocks, in turn, intrude and enclose masses of the mafic dike rock. Some of these felsic stringers are pegmatitic and consist of plagioclase, orthoclase, quartz, and biotite; they appear to be late-stage differentiates of the tonalitic rocks. Locally, the mafic dikes show offsets, whereas the immediately enclosing tonalitic rock shows no fault. Thus, the mafic dikes must have been solid and rigid in a still partly molten tonalitic mass. The presence of subhedral, apparently magmatic epidote in the tonalite suggests crystallization at depths greater than about 25 kilometers (Zen and Hammarstrom, 1984; Zen, 1985).

Continue west on U.S. 12 through the Kamiah plutonic complex to Orofino. Mafic dikes are well represented between Stop 7 and milepost 61.8.

44 Bridge across Clearwater River at Orofino.

0.0 Reset odometer at Orofino bridge.

Cross the Clearwater River at Orofino and abruptly turn west on Idaho Highway 7 toward Ahsahka. Mylonitized Kamiah plutonic complex rocks are exposed along the road.

4.0 Turn north (right) in Ahsahka toward the base of Dworshak Dam, at the first road past (west of) the bridge, and follow the west bank of the North Fork of the Clearwater River.

4.2 **Stop 8: Mylonite of the western Idaho suture zone.**

Stop in the paved parking area under the powerlines.

This exposure shows the synplutonic mafic dikes deformed nearly parallel to the mylonitic foliation and demonstrates the sense of shear by the offset of late pegmatitic veins where they cross the mafic dikes. Quartz diorite/tonalite of the Kamiah plutonic complex is deformed into a strongly foliated and lineated mylonitic gneiss. The rock was completely recrystallized during high-temperature deformation to form hornblende-plagioclase-biotite-quartz mylonitic gneiss. The average orientation of the foliation is S. 64° E. / 55° NE., and the lineation plunges 55° to N.55°E., the direction of mylonitic movement. Both have probably been steepened by movement on late, southwest-dipping, high-angle reverse shear zones.

Synplutonic fine-grained (chilled in the tonalitic magma) mafic dikes outside the shear zone (e.g., at Stop 7) here have been transposed by deformation to be nearly parallel with the foliation. The mafic dikes have been deformed, apparently by simple shear, from a steep orientation (as seen outside the shear zone) into their present concordance, within 1 degree of the foliation; the necessary shear strain requires about 85 kilometers of displacement parallel to the mylonitic lineation (Strayer and others, 1987; Strayer, 1988). The sense of displacement is shown by several features. Late-stage pegmatitic dike differentiates of the tonalitic body range from nearly undeformed to veins and augen trains completely transposed parallel to the foliation. Pegmatitic dikes with intermediate amounts of deformation cross the foliation in the quartz diorite/tonalite and are smeared out more nearly parallel to the foliation where they cross the apparently less competent mafic dikes. Their offset clearly demonstrates a top-to-the-southwest sense of movement. Rare mica "fish" with the asymmetry of a type II S-C mylonite (cf. Lister and Snoke, 1984) show the same sense of movement (Strayer, 1988). Local isoclinal folds deform the transposed mafic layers, pegmatites, and their enclosing mylonitic foliation. Such deformation is consistent with the large shear strains suggested above.

Return to Ahsahka. Turn right (west). Proceed about 0.6 mile as far as the sign indicating the road to Dworshak Dam. Turn right again. Proceed to the dam and cross the dam. Turn right (south) to the maintenance offices to obtain permission to enter the quarry. Then angle left (0.4 mile southeast of dam) past the locked gate on the gravel road. Continue less than 0.5 mile up to the quarry.

Stop 9: Mylonite quarry at Dworshak Dam.

Excavation of coarse rock fill for the dam has provided an outstanding three dimensional view of the mylonite. The longer side walls of the quarry have been cut parallel to the mylonitic lineation; the shorter end walls are perpendicular to the trend of the lineation. The long southeast wall of the quarry (parallel the lineation and to the plane of symmetry of the overall fabric) shows very gentle open waves in the nearly planar foliation with the mafic layers completely transposed into parallelism with the foliation. The late-stage pegmatites show the same sense of offset, top-to-the-southwest, with preferential strain in the less competent, fine-grained mafic layers as at Stop 8. Viewed down the plunge of the lineation, in the northeast wall of the quarry, the pegmatites are poorly oriented. The overall macrosymmetry of the fabric is therefore monoclinic, and the overall shear recorded by the pegmatites is dip-slip. It is clear, therefore, that the deformation in this mylonite zone is dominated by simple shear with movement of continental North America (exposed less than 100 meters northeast of the quarry) southwestward over the more-mafic rocks of the Kamiah plutonic complex.

The late-stage pegmatitic differentiates of the quartz diorite/tonalite form during final crystallization of the pluton at high temperatures and pressures. That they range from completely transposed to undeformed demonstrates that shear strain was active as they began to form but ceased before injection of the latest pegmatites. Thus, the latest stages of crystallization of the dikes record the high temperature and time of completion of deformation in this part of the western Idaho suture zone. The youngest argon-argon dates on the Kamiah plutonic complex (Snee and others, 1987) are about 82 Ma, suggesting that crystallization, deformation, and uplift of the complex were complete by that time.

ACKNOWLEDGEMENTS

I thank Bill Bonnichsen of the Idaho Geological Survey for his careful review of the manuscript.

REFERENCES

Armstrong, R. L., W. H. Taubeneck, and P. O. Hales, 1977, Rb-Sr and K-Ar geochronometry of Mesozoic granitic rocks and their Sr isotopic composition, Oregon, Washington, and Idaho: Geological Society of America Bulletin, v. 88, p. 397-441.

Chase, R. L., 1973, Petrology of the northeast border zone of the Idaho batholith, Bitterroot Range, Montana: Montana Bureau of Mines and Geology Memoir 43, 28 p.

Chase, R. L., and B. R. Johnson, 1977, Border-zone relationships of the northern Idaho batholith: Northwest Geology, v. 6, no. 1, p. 38-50.

Childs, J. F., 1982, Geology of the Precambrian Belt Supergroup and the northern margin of the Idaho batholith, Clearwater County, Idaho: University of California, Santa Cruz, Ph.D. dissertation, 491 p..

Fleck, R. J., and R. E. Criss, 1985, Strontium and oxygen isotope variations in Mesozoic and Tertiary plutons of central Idaho: Contributions to Mineralogy and Petrology, v. 90, p. 291-308.

Foster, D. A., and D. W. Hyndman, in press, Magma mixing between synplutonic mafic dikes and granite, Idaho batholith, central Idaho and western Montana, *in* J. L. Anderson, editor, The Nature and Origin of Cordilleran Magmatism: Geological Society of America Memoir 174.

Greenwood, W. R., and D. A. Morrison, 1973, Reconnaissance geology of the Selway-Bitterroot Wilderness Area: Idaho Bureau Mines and Geology Pamphlet 154, 30 p.

Hietanen, Anna, 1961, Relationship between deformation, metamorphism, metasomatism, and intrusion along the northwest border zone of the Idaho batholith,

Idaho: U.S. Geological Survey Professional Paper 424-D, p. 161-164.

———, 1962, Metasomatic metamorphism in western Clearwater County, Idaho: U.S. Geological Survey Professional Paper 344-A, 116 p.

Hyndman, D. W., 1981, Controls on source and depth of emplacement of granitic magma: Geology, v. 9, p. 244-249.

———, 1983, The Idaho batholith and associated plutons, Idaho and western Montana, *in* J. A. Roddick, editor, Circum-Pacific Batholiths: Geological Society of America Memoir 159, p. 213-240.

———, 1984, A petrographic and chemical section through the northern Idaho batholith: Journal of Geology, v. 92, p. 83-102.

Hyndman, D. W., David Alt, and J. W. Sears, 1988a, Post-Archean metamorphic and tectonic evolution of western Montana and northern Idaho, *in* W. G. Ernst, editor, Metamorphism and Crustal Evolution of the Western United States: Prentice Hall, Englewood Cliffs, N.J., p. 332-361.

Hyndman, D. W., and D. A. Foster, 1988, The role of tonalites and mafic dikes in the generation of the Idaho batholith: Journal of Geology, v. 96, p. 31-46.

Hyndman, D. W., L. M. Strayer, and J. W. Sears, 1988b, The western Idaho suture zone II: Implications from the Kamiah pluton: Geological Society of America Abstracts with Programs, v. 20, no. 6, p. 422.

Jens, J. C., 1974, A layered ultramafic intrusion near Lolo Pass, Idaho: Northwest Geology, v. 3, p. 38-46.

Lister, G. S., and A. W. Snoke, 1984, S-C mylonites: Journal of Structural Geology, v. 6, no. 6, p. 617-638.

Lund, Karen, L. W. Snee, and J. F. Sutter, 1985, Style and timing of suture-related deformation in island arc rocks of western Idaho: Geological Society of America Abstracts with Programs, v. 17, no. 6, p. 367.

Nold, J. L., 1974, Geology of the northeastern border zone of the Idaho batholith: Northwest Geology, v. 3, p. 47-52.

Norwick, S. A., 1972, The regional Precambrian metamorphic facies of the Prichard Formation of western Montana and northern Idaho: University of Montana Ph.D. dissertation, 159 p.

Reid, R. R., 1987, Structural geology and petrology of a part of the Bitterroot lobe of the Idaho batholith, *in* T. L. Vallier and H. C. Brooks, editors, Geology of the Blue Mountains Region of Oregon, Idaho, and Washington: The Idaho Batholith and Its Border Zone: U.S. Geological Survey Professional Paper 1436, p. 37-58.

Reid, R. R., Enid Bittner, W. R. Greenwood, Steve Ludington, Karen Lund, W. E. Motzer, and Margo Toth, 1979, Geologic section and roadlog across the Idaho batholith: Idaho Bureau Mines and Geology Information Circular 34, 20 p.

Reid, R. R., D. A. Morrison, and W. R. Greenwood, 1973, The Clearwater orogenic zone: A relic of Proterozoic orogeny in central and northern Idaho: Belt Symposium, University of Idaho, Moscow, p. 10-56.

Rice, J. M., 1987, Polymetamorphism in northern Idaho: A schistose scenario of Belt and batholith: Geological Society of America Abstracts with Programs, v. 19, no. 7, p. 818.

Ross, D. C., 1985, Mafic gneiss complex (batholith root?) in the southernmost Sierra Nevada, California: Geology, v. 13, p. 288-291.

Schuster, R. D., and M. E. Bickford, 1985, Chemical and isotopic evidence for the petrogenesis of the northeastern Idaho batholith: Journal of Geology, v. 93, p. 727-742.

Simpson, S. J., 1985, Structure and petrology of the Rhodes Peak cauldron, Montana and Idaho: University of Montana M.S. thesis, 105 p.

Snee, Larry, Karen Lund, and Gary Davidson, 1987, Ages of metamorphism, deformation, and cooling of juxtaposed oceanic and continental rocks near Orofino, Idaho: Geological Society of America Abstracts with Programs, v. 19, p. 335.

Strayer, L. M., IV, 1988, Field guide to deformation and sense of displacement in mylonitic rocks near Orofino, Idaho, *in* S. E. Lewis and R. B. Berg, editors, Precambrian and Mesozoic Plate Margins: Montana, Idaho, and Wyoming, with Field Guides for the 8th International Conference on Basement Tectonics: Montana Bureau of Mines and Geology Special Publication 96, p. 165-169.

Strayer, L. M., IV, D. W. Hyndman, and J. W. Sears, 1987, Movement direction and displacement estimate in the western Idaho suture zone mylonite: Dworshak Dam/Orofino area, west central Idaho: Geological

Society of America Abstracts with Programs, v. 19, no. 7, p. 857.

Taubeneck, W. H., 1971, Idaho batholith and its southern extension: Geological Society of America Bulletin, v. 82, p. 1899-1928.

Toth, M. I., 1987, Petrology and origin of the Bitterroot lobe of the Idaho batholith, *in* T. L. Vallier and H. C. Brooks, editors, Geology of the Blue Mountains Region of Oregon, Idaho, and Washington: The Idaho Batholith and Its Border Zone: U.S. Geological Survey Professional Paper 1436, p. 9-35.

Wehrenberg, J. P., 1972, Geology of the Lolo Peak area, northern Bitterroot Range, Montana: Northwest Geology, v. 1, p. 25-32.

Whitney, J. A., 1975, The effects of pressure, temperature, and X_{H2O} on phase assemblages in four synthetic rock compositions: Journal of Geology, v. 83, p. 1-81.

Wiswall, Gil, and D. W. Hyndman, 1986, Emplacement of the main plutons of the Idaho batholith, *in* T. L. Vallier and H. C. Brooks, editors, Geology of the Blue Mountains Region of Oregon, Idaho, and Washington: The Idaho Batholith and Its Border Zone: U.S. Geological Survey Professional Paper 1436, p. 59-72.

Zen, E-an, 1985, Implications of magmatic epidote-bearing plutons on crustal evolution in the accreted terranes of northwestern North America: Geology, v. 13, p. 266-269.

Zen, E-an, and J. M. Hammarstrom, 1984, Magmatic epidote and its petrologic significance: Geology, v. 12, p. 515-518.

Transect Through the Baker and Wallowa-Seven Devils Terranes, Northeastern Oregon

Ellen Morris Bishop[1]
M. Allan Kays[2]

INTRODUCTION

This field trip guide traverses the southeastern part of the Baker terrane and the southern part of the Wallowa-Seven Devils terrane in northeastern Oregon. The rocks examined are mostly south of Baker, Oregon, along the Snake River and in the southern Wallowa Mountains. The authors have drawn heavily from previous geological studies of the area (Ashley, 1966; Brooks, 1979; Brooks and others, 1976; Brooks and Vallier, 1978; Vallier, 1977; Vallier and others, 1977; Prostka, 1962; Taubeneck, 1964) and the unpublished maps and reports of the University of Oregon field geology classes, 1981-1988. In the following we provide (1) descriptions of the geologic units and brief summaries of their stratigraphic, structural, and petrologic relations and (2) a road log for the field trip.

[1] *Department of Geology, Oregon State University, Corvallis, Oregon 97331.*
[2] *Department of Geological Sciences, University of Oregon, Eugene, Oregon 97403.*

CHARACTER OF THE TERRANES AND THEIR UNITS

Terranes of the Blue Mountains

The recognition that the Cordillera of western North America consists in large part of unrelated ensimatic crustal fragments with significant latitudinal displacements has led to the concept of continental growth by peripheral accretion (Jones and others, 1977; Coney and others, 1980; Saleeby, 1983). Inherent to the concept is the notion of fault-bounded lithotectonic units composed of separate stratigraphic sequences that record geologic history different from adjacent units. Units have distinctive and identifiable lithologic characteristics, but inter-

Bishop, E. M., and M. A. Kays, 1989, Transect through the Baker and Wallowa-Seven Devils terranes, northeastern Oregon, *in* V. E. Chamberlain, R. M. Breckenridge, and Bill Bonnichsen, editors, Guidebook to the Geology of Northern and Western Idaho and Surrounding Area: Idaho Geological Survey Bulletin 28, p. 65-84.

nal stratigraphic order is limited. Recognizing that these lithotectonic units are not formations in the classic sense, Irwin (1972) applied the term "terrane" to such units in the Klamath Mountains. The Blue Mountains of eastern Oregon contains an assemblage of tectonically emplaced and juxtaposed Paleozoic and Mesozoic terranes. The regional terranes that we are concerned with here have been named by Silberling and others (1984) and include the Baker terrane, the Wallowa or Wallowa-Seven Devils arc terrane, and the Olds Ferry terrane (Figure 1).

The Baker terrane consists of a variety of tectonically juxtaposed rocks including chert, argillite, graywacke, conglomerate, limestone (Tethyan and North American), greenstone, gabbro and peridotite, amphibolite, high-pressure schist, and serpentinite. Permian and Late Triassic ages are well documented for parts of the terrane as we discuss below. All parts of the terrane are metamorphosed, although grade and deformational character vary considerably across its extent. Despite internal disruption, the lithologic integrity of large parts of the Baker terrane is preserved; the structural condition is probably consistent with the description of "broken formations" (Hsu, 1968). In other parts, more competent lithologies are isolated in a matrix of serpentinite and are, by definition, melanges (Hsu, 1968). This combination of characteristics, which includes broken formation and melange, undoubtedly led Dickinson (1979) to refer to the assemblage as the "central melange terrane." Saleeby, (1983) in an attempt to correlate terranes of the North American Cordillera on the basis of lithologic, faunal, and tectonic affinities, identifies the Baker terrane to be of "Cache Creek affinity." Other terranes of Cache Creek affinity with similar plutonic and metamorphic lithologies and ages are the San Juan terrane in northwest Washington and the North Fork, Fort Jones, and perhaps eastern Hayfork terranes in the Klamath Mountains (Kays and others, 1988).

The Wallowa-Seven Devils terrane consists of (1) intermediate to gabbroic plutonic rocks (subarc?) of low metamorphic grade and Permian to Triassic age (Walker, 1979), (2) basaltic-andesitic-dacitic volcanic and volcaniclastic rocks of similar age, and (3) overlying, unconformable Late Triassic to Early Jurassic platform carbonates and fine-grained clastic rocks. These rocks are intruded by nearly undeformed, tonalitic to granodioritic plutons of the Late Jurassic Wallowa batholith. The Wallowa terrane has been correlated with Wrangellia by Jones and others (1977). However, Sarewitz (1983) points out the uncertainties in such correlation on the basis of lithologic, geochemical, and age differences. As synthesized by Saleeby (1983), Wrangellia is part of an even larger collage termed "Wrangellian super terrane," which consists of Wrangellia, the Alexander and Peninsular terranes in coastal Alaska and British Columbia, Jurassic-Cretaceous overlap sequences, and cross-cutting plutons. Wrangellia proper consists of Carboniferous primitive arc and subarc intrusive rocks, which are separated from Triassic plateau pelagic rocks, carbonates, and flood basalts by an unconformity. The Alexander terrane has similar characteristics but also includes metamorphosed basement arc rocks of probable Ordovician age or older.

The Olds Ferry terrane (Figure 1) lies south of the Wallowa-Seven Devils terrane and easternmost Baker terrane and extends northeastward into Idaho as far as Cuddy Mountain (Brooks, 1979). It is dominated by Late Triassic volcanic, volcaniclastic, and epiclastic rocks similar in composition to those of the Wallowa terrane. The volcanic sequence, named the Huntington Formation (Brooks and others, 1976), is generally younger than metavolcanic units of the Wallowa terrane. It may be equivalent in age to post-arc volcanic rocks of the Lower Sedimentary Series in the southern Wallowa Mountains. Minor amounts of limestone with a large content of volcanic debris are interbedded with siltstone. Jurassic flysch rocks of the Weatherby Formation unconformably overlie the deformed, low-grade, Late Triassic volcanic rocks and are in high-angle, thrust-faulted contact with the Burnt River Schist of the Baker terrane along the Conner Creek fault. The flysch units have unusually well-developed cleavage, and the rocks are locally phyllitic approaching the Conner Creek fault. The relation of the Olds Ferry terrane to the older, somewhat more mafic Wallowa-Seven Devils rocks to the north is uncertain. Brooks (1979) suggests on the basis of age and lithologic distinctions that the Olds Ferry terrane represents a different arc assemblage.

Geologic Units of the Baker Terrane

The Baker terrane is dominated by two units—the Burnt River Schist and the Elkhorn Ridge Argillite. Both are mostly cherts and siliceous argillites with varying amounts of mafic volcanic greenstones intercalated with, or enclosed by, the sedimentary rocks. Plutonic greenstone bodies, serpentinites, and the coherent ophiolites are mapped separately but seem to be an integral part of the terrane. The Elkhorn Ridge Argillite occupies the northern and northwestern part of the terrane; Burnt River Schist occurs in the south and southeast (Figure 2). Westward across the Baker terrane, serpentinite-matrix melanges that contain schists and amphibolites, in addition to the other Baker terrane lithologies just listed, increasingly displace deformed, but relatively coherent, sedimentary/volcanic sequences. Thus, the Baker terrane can be subdivided into three

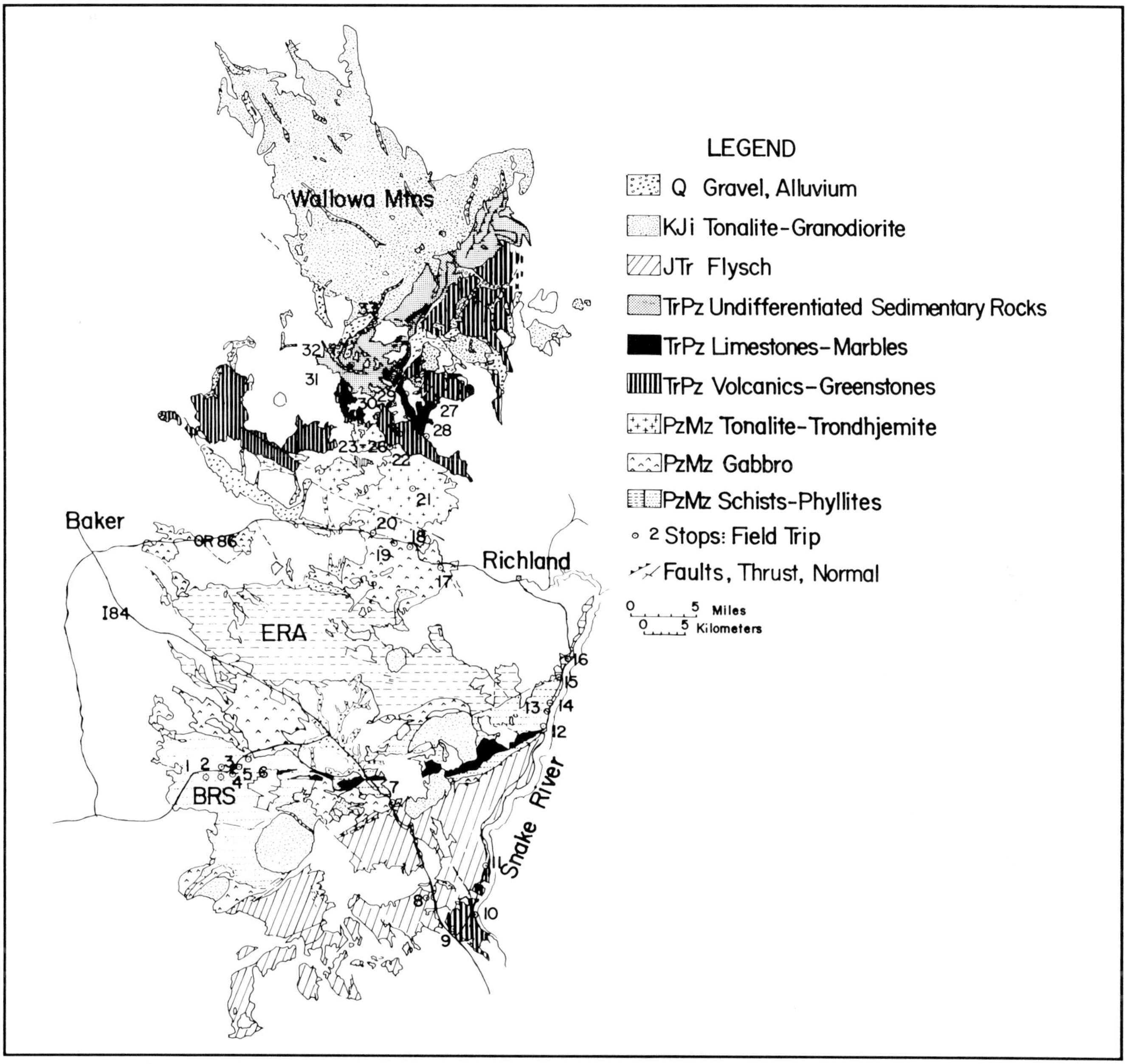

Figure 1. Index map of northeast Oregon showing terranes and localities discussed in the text. Jurassic-Cretaceous batholiths are omitted. The Olds Ferry terrane is shown as "arc terrane" but has a "forearc basin terrane" cover (Weatherby Formation) in the vicinity of Huntington and Ironside. The Wallowa-Seven Devils arc terrane is mainly north of Sparta and Baker. The "forearc basin terrane" south of John Day is known as the Izee terrane. The problematical ophiolitic gabbro-peridotite complexes are the Canyon Mountain complex south of John Day and the Sparta complex in the vicinity of Sparta.

parts: (1) Elkhorn Ridge Argillite-dominated, (2) Burnt River Schist-dominated, and (3) serpentinite melange-dominated parts (Kays, Ferns, and Brooks, 1988). Because of the similarities in their diversity, the igneous components of the Elkhorn Ridge Argillite, Burnt River Schist, and serpentinite melange subdivisions of Baker terrane volcanic and plutonic components are discussed separately.

Elkhorn Ridge Argillite: Sedimentary Units

Elkhorn Ridge Argillite forms the bulk of sedimentary exposures within the Baker terrane. The unit was named for exposures on Elkhorn Ridge in the Sumpter quadrangle by Gilluly (1937). Sedimentary lithologies include red, green, tan, and rare black cherts, rather dark argillites and minor interlayered coarse-grained clastic and volcaniclastic conglomerates, grits, and sandstones with

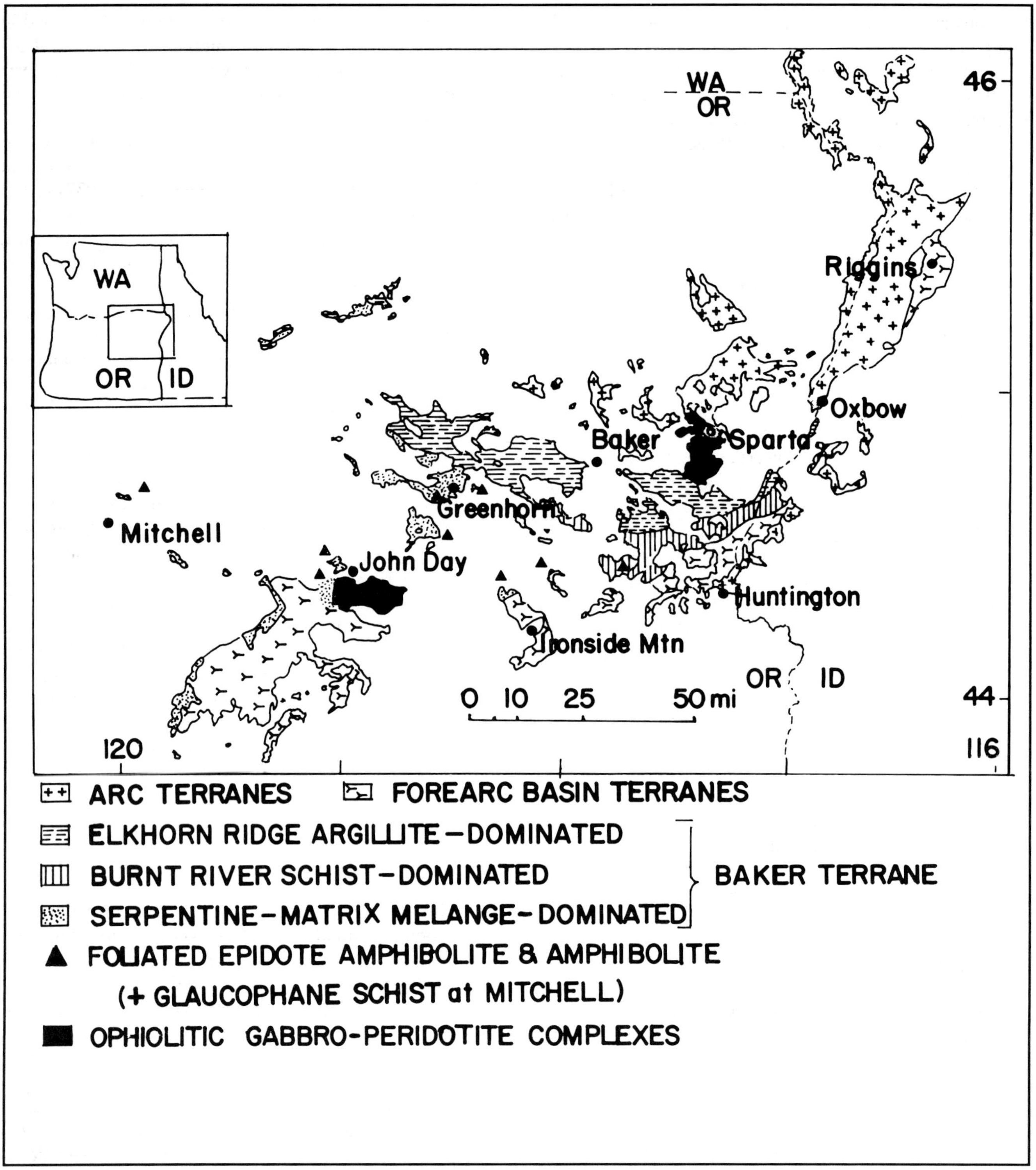

Figure 2. Simplified geologic map for the area of the field guide. Numbers and circles refer to stops for the field trip. The symbols in the legend are: "JTr Flysch": Jurassic Weatherby Formation; "TrPz Undifferentiated Sedimentary rocks": mainly Hurwal Formation in the southern Wallowa Mountains; "TrPz Limestones-Marbles": mainly Martin Bridge Limestone in the southern Wallowa Mountains, and Nelson Limestone in "PzMz Schists-Phyllites" in the southern part of the map; "TrPz Volcanics-Greenstones": mainly Clover Creek Greenstone and Gold Creek Greenstone in the southern Wallowa Mountains, and Huntington Formation along the Snake River; "PzMz Schists-Phyllites": Elkhorn Ridge Argillite designated "ERA" on the map and Burnt River Schist designated "BRS" on the map.

siliceous matrixes. Also present are minor, but locally mappable, discontinuous beds, lenses or lentils of limestone and marble. These bodies have generally elongate, deformed, mapped expression to several hundreds of meters, but smaller bodies are less than a meter long. The limestone and marble bodies are generally enclosed within argillite.

The pelitic phyllites are composed mainly of quartz-muscovite-chlorite, smaller amounts of albite-biotite(locally)-graphite(?)-clinozoisite, and subordinate calcite. Quartzites may locally contain spessartine, and quartzo-feldspathic phyllites have albite as a major mineral. The marbles have, in addition to calcite, subordinate garnet, muscovite, and quartz. Phyllosilicates define penetrative (S1) schistosity of generation D1 in these rocks. The schistosity-cleavage appears to be similar to that described by Ashley (1966) for the Burnt River Schist (i.e., axial planar to D1 generation folds with F1 axes). The schistosity is deformed by crenulation folds of a later generation (D2?). A secondary cleavage (S2?) approximately axial planar to crenulation folds is of variable strength throughout the region.

Conodont faunas from cherts of the Baker terrane yield ages that range from Late Permian (Guadalupian) to Late Triassic (Norian) or Early Jurassic (Pleinsbachian or Toarcian) (Blome and others, 1986). Conodont and fusilinid faunas in limestones yield a range of ages that according to Morris and Wardlaw (1986) are Middle to Late Devonian, Middle to Late Pennsylvanian, Permian, and Middle to Late Triassic. The Permian and Triassic fusulinid faunas have been identified as having both North American and Tethyan affinities (Morris and Wardlaw, 1986). The age diversity of the limestones, and their occurrence as lentils or blocks, especially those with tectonic contacts with enclosing argillites, suggest tectonic mixing. For example, Morris and Wardlaw (1986) suggest that arc-related North American blocks became mixed with oceanic Tethyan blocks, perhaps during the pre-accretionary amalgamation (?) event that formed the Baker terrane. Pessagno and Blome (1986), on the basis of faunal characteristics, suggest northward movement from Permian and Early Triassic ("Tethyan") latitudes (18 degrees) to boreal latitudes (30 degrees) by later in the Jurassic.

Burnt River Schist: Sedimentary Units

The Burnt River Schist in the southeast part of the Baker terrane is composed of mostly sheared, deformed argillites and cherts that are similar to those in the Elkhorn Ridge Argillite. Relations between these two units are uncertain because the contact is not well exposed. Most workers, however, consider the Burnt River Schist to be a more deformed equivalent of the Elkhorn Ridge Argillite, i.e., small-scale structures such as penetrative foliation-schistosity are better developed in the Burnt River Schist. The metamorphic mineral assemblages in metasedimentary rocks of the Burnt River Schist are essentially the same as those of the Elkhorn Ridge Argillite.

Excellent descriptions of the major units of the Burnt River Schist are given by Ashley (1966) for exposures in Burnt River canyon, designated by Gilluly (1937) as the type locality, 30 kilometers southeast of Baker, Oregon. In this area, pelitic phyllite dominates but greenstones and greenschists (stops 3 and 1, respectively, Figure 2) are nearly as abundant and locally interlayered with metasedimentary rocks and each other. Other lithologies include phyllitic quartzo-feldspathic metapsammites and metaconglomerates. Pelitic quartzites and metacherts are widespread. Podlike bodies and scattered lenses of limestone and marble (stop 4, Figure 2) are a few inches to about one kilometer long and generally aligned parallel to the foliation in the adjacent phyllite. Despite their locally strong deformation and recrystallization, silicification and dolomitization, many marble bodies are fossiliferous. The limestone bodies mapped by Ashley (1966) are folded, and their mapped expression appears to define first generation (F1) folds in the Burnt River Schist. Ashley relates such axial planar cleavages in phyllites and limestones to D1 deformation. The folds and associated cleavage-foliation recognizable in the limestones suggests that Ashley was justified in assigning chronologic significance to D1 and D2 deformational features in the Burnt River Schist which appear to be similar to those in the Elkhorn Ridge Argillite as described above.

Volcanic Greenstones

Massive, recrystallized greenstones with relict volcanic textures and minerals are widespread throughout the Burnt River Schist and Elkhorn Ridge Argillite and have been mapped as parts of those units. Although metabasaltic rocks are well represented and probably dominate greenstone compositional types, more silicic metavolcanic rocks of andesitic composition are also present. The careful mapping of Ashley (1966) probably best illustrates the occurrences of these rocks. Greenstones and greenschists from Burnt River canyon contain dominant chlorite-albite-pumpellyite and lesser muscovite-sphene-actinolite and minor calcite-clinozoisite-quartz (Ashley, 1966). Some metabasaltic assemblages contain relict volcanic clinopyroxene (e.g., stop 14, Figure 2) of a range of compositions (Bishop, 1988). Greenstones, which locally have pillow structure (e.g., stop 16, Figure 2), have poorly developed cleavage-schistosity compared with adjacent metasedimentary units. However, where these fabric features are present in greenschists and greenstones, they are defined similarly to and parallel with those of the metasedimentary rocks. Phyllosilicates, dominantly chlorite, are more abundant and well aligned in the metamorphosed equivalents of pyroclastic and volcaniclastic rocks. Actinolite, which may be widespread locally, is rarely well aligned. Ashley (1966) defines schistosity (S1) in the greenstones and greenschists as equivalent to D1 deformation in the phyllites. In these rocks D2 deformation is generally not as well defined as it is in the adjacent metasedimentary units.

A reconnaisance study of major and trace element data of basaltic greenstones and their relict clinopyroxenes from several locations (Figures 3 and 4) appear to reflect a variety of plate tectonic environments (Bishop, 1988). Note in particular that clinopyroxenes from the Burnt River Schist plot in the "within plate" alkalic basalt field and are enriched in TiO_2 in comparison with greenstones from other units. Overall, the character and compositions of Baker terrane greenstones and their relict clinopyroxenes are similar to those described for the Permian to Late Triassic pillow basalts and other mafic metavolcanic rocks of the North Fork terrane (Ando and others, 1983) in the Klamath Mountains.

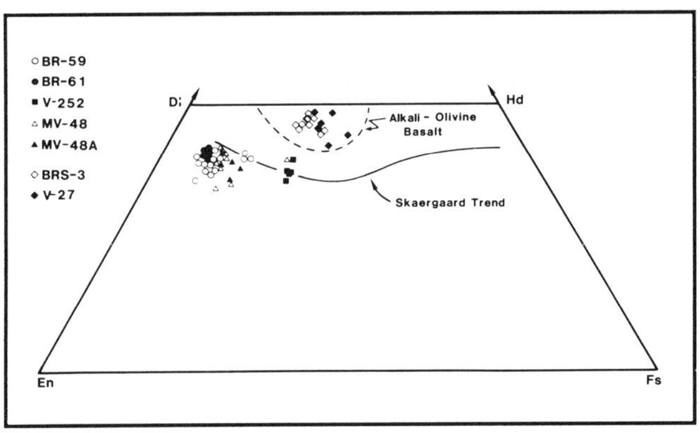

Figure 3. Pyroxene quadrilateral for clinopyroxenes in the greenstone metavolcanic rocks of the Baker terrane. Compositions are in terms of diopside (Di), hedenbergite (Hd), enstatite (En), and ferrosilite (Fs) end member minerals. The symbols with numbers refer to sample numbers.

Plutonic Greenstones and Peridotite-Gabbro Complexes

Basic and intermediate plutonic rocks are widespread and together form an important lithologic group in the Baker terrane. For convenience during this discussion, we divide the occurrences into two groups: (1) plutonic complexes located near the northern and southern margins of the Baker terrane and having recognizable vertical "stratigraphy," i.e., with peridotites at the base followed upward by layered and massive gabbro, diorite, quartz diorite, and albite granite or trondhjemite; (2) the same rock types as the complexes, but with no recognizable order, dominated by sheared gabbro widespread throughout the Baker terrane.

U-Pb zircon ages of a number of these basic to silicic metaplutonic rocks including those of the Sparta and Canyon Mountain complexes are older than about 219 Ma (Walker, 1983, and personal communication). For example, Walker (1983) indicates zircon ages that range from 268 to 262 Ma for quartz diorites and albite granites from the Canyon Mountain complex. Gerlach and others (1988) report $^{40}Ar/^{39}Ar$ ages of 261 and 262 Ma for late diorites, and a Nd isochron of 267-274 Ma for cumulate and isotropic gabbro and ultramafic rocks in the Canyon Mountain complex. Ave'Lallemant and others (1980) report $^{40}Ar/^{39}Ar$ plateau ages of 223 Ma on hornblendes of quartz diorite and trondhjemite in the Sparta complex. The occurrence and metamorphism of the sheared gabbros-diorites scattered throughout the Baker terrane suggest that they are older than the regional, low-grade Norian recrystallization (Ave'Lallemant and others, 1980) that affected the associated metavolcanic and metasedimentary rocks.

The mapping of Prostka (1962) indicates that the Sparta complex (Figure 1) consists of isotropic or unlayered and layered gabbros (e.g., Stops 18 and 19, respectively, Figure 2) that contain serpentinized peridotite in the southern part of the complex. The gabbro is followed northward by quartz diorite and a belt of trondhjemite that abuts the Clover Creek Greenstone of the adjacent Wallowa terrane. The peridotite, which represents no more than about 10 percent of the complex, has uncertain contact relations with the partly sheared gabbro that surrounds it. The quartz diorite and trondhjemite (Stops 20 and 21, respectively, Figure 2) are intruded by (coeval?) diabase and keratophyre dikes, but it is uncertain whether they are related to dikes and flows of similar composition in the overlying Clover Creek Greenstone. Phelps and Ave'Lallemant (1980) suggest, on the basis of the regional geologic setting and the major and trace element compositions of the rocks in the Sparta complex, that it is the plutonic equivalent to the low K_2O, tholeiitic island arc magmatic series.

The Canyon Mountain complex is an east-west-trending, 4-kilometer-thick Permian ophiolitic sequence with north to south order of peridotite, pyroxenite, gabbro, diorite, plagiogranite, diabase, and keratophyre exposed over more than 150 square kilometers south of John Day, Oregon (Figure 1). On the basis of chemistry and lithologies, previous workers have considered that the Canyon Mountain complex is anomalous in comparison to other ophiolites (Thayer, 1977; Himmelberg and Loney, 1980; Ave'Lallemant, 1976; Gerlach and others, 1981; Mullen, 1985). For example, the complex is divided into dissimilar eastern

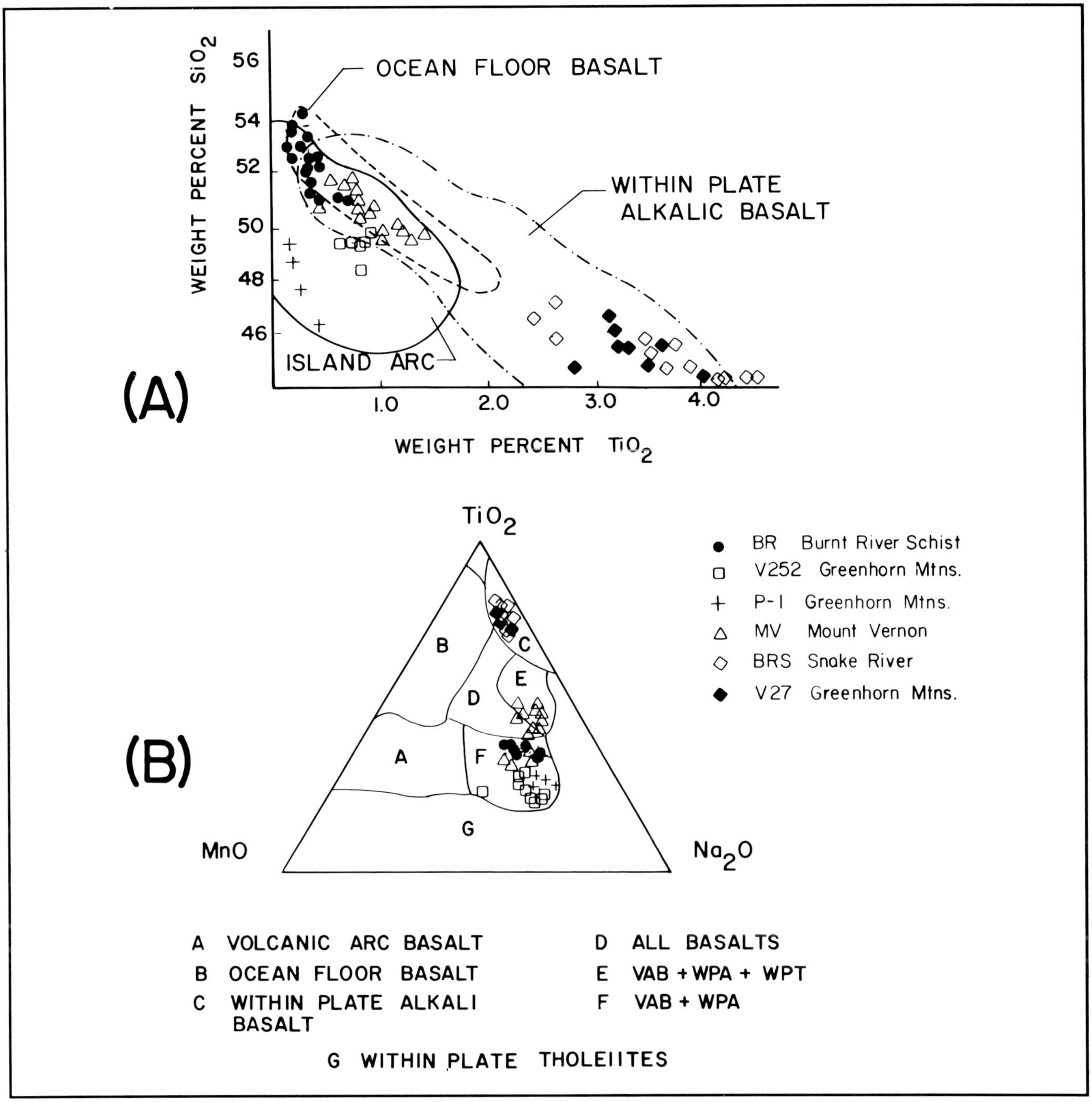

Figure 4. (A) SiO_2-TiO_2 plot for clinopyroxenes from northeast Oregon greenstones. Symbols refer to sample numbers as in Figure 3. (B) MnO-TiO_2-Na_2O discriminant diagram for clinopyroxenes in basaltic greenstone samples in (A).

and western parts by a northeast-trending fault, and there is intrusion of slightly fractionated gabbro into tectonite peridotite along the boundary between the two parts. The complex has no overlying basaltic pillow lava sequence, and the diabase sheeted dike complex is parallel rather than normal to the strike of the adjacent lithologic units. Bishop (1988) has shown that the cumulate gabbros are more calcic (Figure 5) than in many other ophiolite complexes and plot in alkalic to calc-alkaline cumulate fields rather than that of mid-ocean ridge basalt (ophiolite) field. Gerlach (1980) and Mullen (1983, 1985) suggest on the basis of petrologic data that the Canyon Mountain complex was associated with island arc or forearc magmatism.

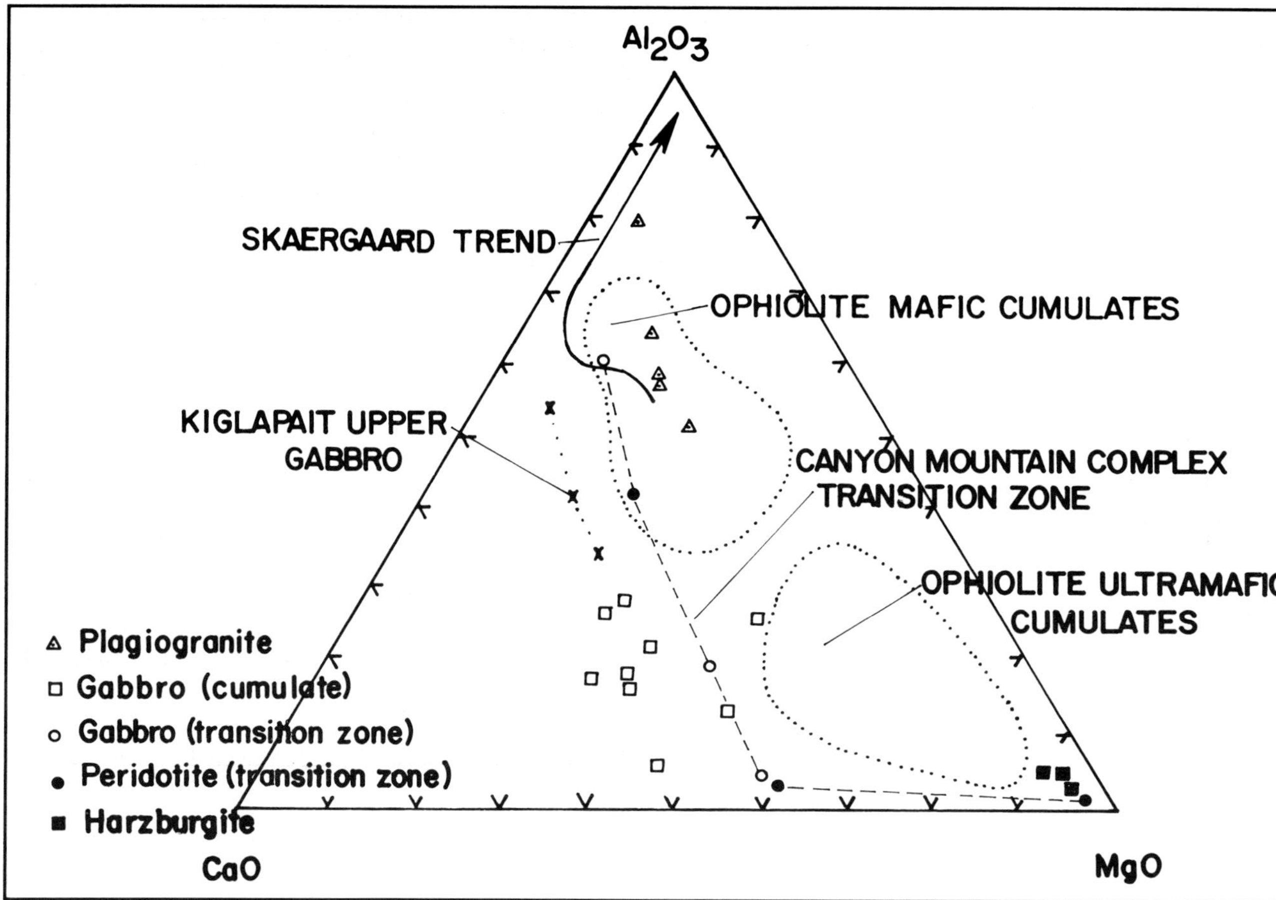

Figure 5. MgO-CaO-Al_2O_3 for the rocks of the Canyon Mountain complex.

The sheared gabbros of category (2), as noted in the first paragraph of this section, have an uncertain emplacement mechanism. Gabbro is probably most abundant, but quartz diorite and albite granite are also present as mappable bodies in close or (more rarely) ordered association with the gabbros (Brooks and others, 1976). Although previous workers have focused attention on the larger complexes, it seems important to point out that the petrologic nature of the areally significant sheared gabbros and more silicic rocks of the Baker terrane has not been studied independent of their metamorphism. The gabbroic and intermediate rocks are largely recrystallized, but similarly to the basaltic greenstones, relict plutonic minerals are recognizable. The metamorphic assemblage in quartz-free rocks is albite-clinozoisite-epidote-chlorite-prehnite-serpentine ± actinolite ± calcite. Although locally schistose, the metamorphic textures are commonly those of cataclastic microbreccias, protomylonites, and mylonitic gneisses. Thus their assemblages, structures-fabrics, and ages suggest that they are part of an association of rocks that is pre-Norian. Serpentinized peridotites-pyroxenites occur mainly as separate sheared bodies and in some areas dominate the terrane as we describe below.

Serpentinite-Matrix Melanges

The exposed western parts of the Baker terrane are dominated by serpentinite-matrix melanges such as the one around Greenhorn (Mullen, 1978; Ferns and others, 1983; Brooks and others, 1983) and in the area to the west of John Day and the Canyon Mountain complex (Brown and Thayer, 1966; Figure 1). Clasts in the matrix of serpentinite include all the lithologies discussed above, as well as amphibolite, glaucophane-crossite-bearing schists, and coarse-grained conglomerate in the Greenhorn melange. The conglomerate is of special interest because it has clasts similar to metamorphosed lithologies of the melange. Ferns and others (1983) and Brooks and others (1983) regard the conglomerate as Late Triassic on the basis of fossil evidence, and they suggest that it represents a cover sequence that was incorporated into the underlying melange as deformation continued within the Baker terrane. Clasts in the melange vary in metamorphic grade. For example, some peridotite clasts (rarely) have spinel lherzolite assemblages; the amphibolite clasts are commonly amphibolite facies, and clasts of the argillites-phyllites and greenstones are greenschist and subgreenschist facies similar to those of the Elkhorn Ridge Argillite and Burnt River Schist. The melange matrix is generally low-grade greenschist to subgreenschist facies.

The blue amphibole-bearing schistose rocks of the serpentinite-matrix melanges are of special interest because of the tectonic setting that their presence suggests (i.e., subduction zone). Their lithologic similarity and occurrence as tectonic blocks in areas of up to a few miles long in the melange suggest that they may have derived from in-situ, coherent recrystallized sequences such as the lawsonite-glaucophane-bearing blueschists from Mitchell. The sequence at Mitchell consists of interlayered quartz-rich pelitic and calcareous schists with a metamorphic age of 214-223 Ma by K/Ar and $^{40}Ar/^{39}Ar$ on micas (Hotz and others, 1977). Core and rim compositions of blue amphiboles from Mitchell and several of the melange localities in the Baker terrane have been analyzed by electron microprobe. The extremes of core and rim compositions are reported in Figure 6 in plots of amphibole NaM_4 versus Al^{IV} after Brown (1977). The amphiboles in the assemblage from Mitchell record the highest pressure, about 6.5 kilobars, according to the empirically calibrated plot. The Bennett Creek schists in the Greenhorn melange have blue amphibole core compositions that indicate pressures of more than 6 kilobars,

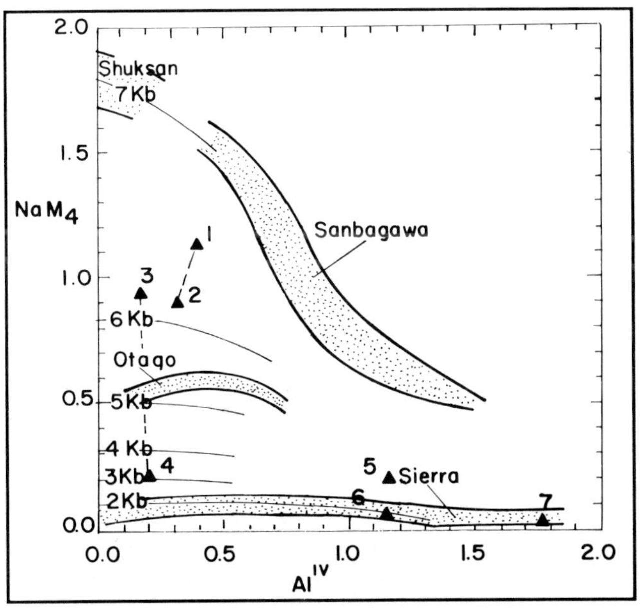

Figure 6. Plot of NaM_4 versus Al^{IV} and pressure for assemblages in well-documented localities associated with amphibole-bearing rocks after Brown, (1977). The amphiboles are from several eastern Oregon localities in the Baker terrane. Samples numbers refer to core and rim compositions from the Mitchell (1,2), Bennett Creek (3,4), and Mine Ridge (5,6) localities, respectively. Rhea Creek numbers (7,8) are separate samples.

but rim compositions indicate pressure between 3 and 4 kilobars. The core to rim progression in time for samples from most locations suggests that all the amphibole-bearing assemblages experienced a change in metamorphic conditions with an attempt at reequilibration to lower pressures and probably higher temperatures (Ernst, 1968; Liou and others, 1975; Brown, 1977). Our work is continuing in order to develop the mineralogical and petrologic details on these interesting assemblages.

Geologic Units of the Wallowa-Seven Devils Terrane

The Wallowa-Seven Devils terrane includes a thick pile of island arc volcanic rocks of Permian and Late Triassic age represented by the Clover Creek Greenstone and Gold Creek Greenstone in the southern Wallowa Mountains (Brooks, 1979; Vallier and Batiza, 1978). The Martin Bridge Limestone and lower part of the Hurwal Formation are younger sedimentary rock sequences of Late Triassic age deposited following arc volcanism. The units of the Wallowa-Seven Devils terrane have been deformed but not disrupted to the same extent as the melange. The Clover Creek Greenstone and Lower Sedimentary Series have low-grade metamorphic assemblages, and all the rock units have been thermally metamorphosed adjacent to the Wallowa batholith. The main part of the terrane traversed in this field trip guide has been mapped and described by Prostka (1962), and we have relied heavily on this work in the following discussion.

Clover Creek and Gold Creek Greenstones

The Clover Creek Greenstone was named by Gilluly (1937) for the sequence of altered volcanic flows and pyroclastic rocks along Clover Creek in the Baker quadrangle where the section is about 4000 feet (1200 m) thick. Although the Clover Creek Greenstone and the overlying Lower Sedimentary Series include Permian and Upper Triassic strata, the strata are all Triassic in the Sparta quadrangle (Brooks and others, 1976). A variety of rock types make up the unit including spilites, several kinds of keratophyres, siliceous tuffs and sedimentary rocks (e.g., Stops 23-26, Figure 2). Spilites along with keratophyres are the most common kinds of volcanic rocks. The spilites occur as flows that are 30 to 100 feet thick. The rocks are dark green, massive, and amygdaloidal. Major minerals are albite, chlorite, epidote, calcite, opaque minerals (oxides and/or sulfides?), and quartz with accessory actinolite, white mica, sphene, leucoxene, and prehnite. Locally, there are relict phenocrysts of pyroxene.

Light greenish gray quartz keratophye is less abundant. This rock type is porphyritic with phenocrysts of albite, quartz, and hornblende in a microcrystalline groundmass of quartz, albite, magnetite and chlorite. Gray, tan, olive, and greenish finely laminated siliceous tuff is dense and brittle, and it breaks with conchoidal fracture. These rocks are composed of broken albite and quartz crystals in a microcrystalline groundmass of quartz + albite ± sericite, chlorite, iron-oxides and sparse shardlike fragments.

A variety of sedimentary rock types occur in the upper part of the unit intercalated with the volcanic rocks. The lithic wackes and conglomerates are composed mainly of fragmental andesitic and dacitic volcanic material. In general, the clasts are poorly rounded and the sorting is poor. Compared to the lithic wackes, volcanic wackes have a greater proportion of crystal fragments and clasts set in a fine-grained chloritic groundmass. The sandstones compared to the wackes have better rounding and sorting. Finely laminated shales and argillites contain finely divided chlorite. There are occasional thin beds of grayish chert. Limestone is also present in apparent faulted association with the greenstone, and it is not clear whether these two rock types were interlayered prior to faulting. According to Prostka (1962), all the rocks are mildly metamorphosed and contain albite, chlorite, prehnite, pumpellyite, epidote, and minor zeolites. The volcanic rocks are more highly altered than the interbedded sedimentary rocks suggesting possible hydrothermal alteration (during submarine eruption?).

Prostka (1962) named the Gold Creek Greenstone for the approximately 2000-foot-thick sequence of flows and breccias near the head of Gold Creek in the Sparta quadrangle. The main distinction recognized by Prostka is that the Gold Creek Greenstone underlies the Lower Sedimentary Series whereas the Clover Creek Greenstone underlies the Martin Bridge Limestone. However, because of rapid facies changes in the upper part of the Clover Creek Greenstone and the lower part of the Lower Sedimentary Series, the contact between those two units is uncertain. For example, the thin clastic units at the top of the Clover Creek Greenstone, as mapped by Prostka (1962), are exposed beneath rock of the Martin Bridge Limestone along Empire Creek. The clastic units may be representative of Lower Sedimentary Series deposition elsewhere.

Lower Sedimentary Series

The Lower Sedimentary Series was described by Smith and Allen (1941) and is considered to be equivalent with the upper part of the Late Triassic Seven Devils Group or Doyle Creek Formation in Hells Canyon (Vallier, 1977). Prostka (1962) separated the series into a lower two-thirds that is dominated by conglomerates and breccias and an upper one-third which consists of distinctive red, purple, and green finer grained sandstones, siltstones, and shales that may grade into similar rocks at the base of the Martin Bridge Limestone, e.g., on the east side of East Eagle Creek on the ridge south of Sullivan Creek. The overall thickness of the unit is probably 1000 to 1500 feet. The lower contact with the Clover Creek Greenstone is not readily determined because sedimentary rocks are found in each unit near the mutual contact. The base has been arbitrarily defined by Follo (1986) as the top of the uppermost major flow unit in the underlying greenstone metavolcanic sequence.

The sequence of coarse clastic rocks above the Clover Creek Greenstone and Gold Creek Greenstone adjacent to main Eagle Creek in Empire Gulch (Stops 24-26 in Figure 2) and up near the head of Paddy Creek (Stop 27, Figure 2), respectively, shows progressive textural and compositional changes upward. Along Empire Gulch, conglomerates and grits mapped at the top of the Clover Creek Greenstone contain abundant angular to subrounded clasts of greenstone, chert, and cherty argillite. The matrix is highly chloritic and also contains angular to subrounded feldspar, quartz, and lithic fragments. In the upper part of the series, poorly sorted conglomerates contain rounded clasts of dacite and andesite porphyry, hornblende and albite granite, chert, and chloritic greenstone. It seems reasonable that the eroded source for the clasts of volcanic and plutonic rocks was the underlying Clover Creek-Gold Creek Greentone and Sparta Complex, respectively. The conglomerates and sandstones are poorly sorted; the clasts occur in an argillaceous matrix identical with argillaceous rocks of the series. The sedimentary rocks display soft-sediment deformation, convolute beds, slump folds, graded beds, and flame structures.

Follo (1986) interprets the sequence of clastic rocks in the Lower Sedimentary Series to be the product of current deposition in relatively shallow submarine fan environments. Such fans have been described by Dickinson (1974) as shallow epiclastic dispersal aprons built up around volcanic archipelagos. In the case of the Lower Sedimentary Series in the southern Wallowa Mountains, the supply of sediment apparently exceeded the subsidence of the basin.

Martin Bridge Limestone

The Martin Bridge Limestone was named by Ross (1938) for the measured section near Martin Bridge by the confluence of Paddy Creek and the main branch of Eagle Creek (near Stop 28, Figure 2). Unfortunately, that section is atypical in comparison with the Martin Bridge Limestone in the northern Wallowa Mountains (see Nolf, 1966). Complex faulting and folding make stratigraphic relations uncertain in the Martin Bridge Limestone and Hurwal Formation, and the two units have not been distinguished by Walker (1979) in the southern Wallowa

Mountains. The University of Oregon compilation map indicates that the contact between dominant limestone (Martin Bridge) and overlying dominant interlayered shale and siltstone (Hurwal) strikes northwest from the confluence of Eagle and East Eagle Creeks toward the upper reaches of Bradley and O'Brien Creeks (near Stop 29, Figure 2).

The Martin Bridge Limestone contains mappable massive and bedded limestones with minor or local unconformities. The Hurwal Formation is dominated by organic-rich siltstones and shales with local lenses and layers of limestone-chert conglomerate beginning near the contact with dominant limestone. This is also the conclusion of Follo (1986). The thickness of the Martin Bridge Limestone according to this distinction varies from less than 100 meters along East Eagle Creek to as much as 400 meters in the section from Empire Creek up Paddy Creek. Our mapping and that of Prostka (1962) indicates further that exposures of Lower Sedimentary Series, Martin Bridge Limestone, and Hurwal Formation as we define them here are structurally controlled. The main structures are northwest-trending broad folds of several miles amplitude that tighten and become complexly overturned northward up East Eagle Creek. The folds are made more complicated by later widespread east-west- to northwest-trending normal faults.

The prominent or readily recognizable lithofacies in the Martin Bridge Limestone are the micritic mudstone carbonates, as recognizable near the type section (Stop 28, Figure 2), and thicker but still well-bedded carbonate grainstones according to Follo (1986). Massive limestone conglomerates are especially abundant near the top of the section or at the base of the overlying Hurwal Formation. Faunal studies of the Martin Bridge Limestone by Smith (1912, 1927) identified Late Triassic corals on Eagle Creek near the type section. Subsequent work by Silberling and Tozer (1968) indicates that the corals are of Lower Norian age in the northern Wallowa Mountains. Stanley and Senowbari-Daryan (1986) describe Norian reefoidal limestones from Summit Point in the southern Wallowa Mountains and suggest that they may be platform carbonates similar to Upper Triassic Dachstein-type reef sequences of the Alps. The coralline limestones near the type section are probably just slumped downslope accumulations and debris sheets of clastic coral fragments (Stanley, 1979).

All limestone lithofacies show progressive development of cleavage and foliation as the limestone changes to marble eastward up East Eagle Creek nearer to the Cornucopia stock and Wallowa batholith. For example, in conglomerates composed of limestone, the clasts are flattened and elongated, the flattening plane corresponding to cleavage in the adjacent argillaceous beds near the confluence of East Eagle Creek and Eagle Creek. In well-bedded sections nearer the contact with the Lower Sedimentary Series along the ridge line south of Sullivan Creek, limestone and argillite beds are transposed by metamorphic cleavage at 30-45 degrees to bedding. Along East Eagle Creek, beginning near the confluence of Little Kettle Creek and continuing northward are well-banded exposures of gray-blue foliated marbles.

The Hurwal Formation

The Hurwal Formation was described by Smith and Allen (1941) as conformable upon the Martin Bridge Limestone in the northern Wallowa Mountains. The Hurwal Formation in the area near the contact with the Wallowa batholith in the southern Wallowa Mountains is limey or siliceous, dominantly fine-grained laminated siltstones-shales with alternating laminae a few millimeters to centimeters thick. The rock is commonly black with abundant organic debris, probably mostly graphite. Pyrite occurs in organic-rich laminae with evidence for primary and secondary crystals, the latter up to a centimeter in diameter grown through the bedding laminae. Limestone conglomerates (Stop 30, Figure 2), sandstones, siltstones, and shales are interlayered near the contact with the Martin Bridge Limestone. However, conglomeratic layers-lenses that contain chert and limestone clasts are common throughout the Hurwal Formation in the southern Wallowa Mountains and especially prominent between Excelsior Gulch and Bennett Peak. Follo (1986) describes the conglomerates of this type as the Excelsior Gulch Member.

As in the underlying Martin Bridge Limestone, the rocks of the Hurwal Formation show the effects of metamorphism and deformation northward along East Eagle Creek and toward the Wallowa batholith. The unit thickness is obviously greatly affected by folding and to a lesser extent by faulting. Follo (1986) estimates a measured thickness of about 500 meters based on exposures in Excelsior Gulch.

Jurassic Granitic Plutons

The Wallowa batholith and the Cornucopia stock are the main plutonic bodies which intrude Triassic rocks in the southern Wallowa Mountains. According to Taubeneck (1987), the Wallowa batholith is a Jurassic composite intrusion emplaced in a mafic-to-felsic sequence that commenced with many small gabbroic bodies. The gabbroic bodies, which contain xenoliths of pyroxene-rich ultramafic rocks, are of mappable dimensions about 2 to 3 miles wide and about 7 miles long concentrated along the southeast margin of the batholith (Stop 32, Figure 2). We also note a peculiar concentration of very narrow dikes, which we term lamprophyre, that have an ultramafic to dioritic range of compositions emanating apparently from the gabbroic centers into the adjacent Hurwal rocks. These rocks as well as the gabbros are intruded and metamorphosed by the main tonalite-granodiorite of the batholith proper. In the southern Wallowa Mountains, the tonalite-granodiorite is dominantly hypidiomorphic granular and coarse-grained and is composed mainly of hornblende-biotite-plagioclase-quartz with lesser microcline. Leucocratic trondhjemitic rocks are especially widespread in the satellitic Cornucopia stock and locally contain magmatic cordierite (Taubeneck, 1964).

The effects of thermal metamorphism by the batholith on the surrounding Triassic rocks are pronounced along the southeast margin. The thermal aureole is about a mile wide in the southeast and grows gradually to about 4 miles wide northeastward along East Eagle Creek at least as far north as the confluence with Curtis Creek. The rocks nearest the contact are hornblende hornfels facies and have slaty cleavage (argillaceous) or banding (marbles-sandstones) consistent with forceful emplacement of the adjacent plutonic rocks which also resulted in strong folding (Taubeneck, 1987).

ROAD LOG

Day 1: Road Log for Baker Terrane Field Guide—Burnt River Canyon and Snake River Canyon

Mileage	Description
0.0	The mileage begins at the U.S. Post Office in Baker. Drive south on Oregon Highway 7.
9.0	Junction of Oregon 7 and Oregon Highway 245. Continue south (left) on Oregon 245.
16.8	This is the approximate summit of Oregon 245 across Dooley Mountain.
18.8	The rocks by the roadside are coarse tuff breccia of the Tertiary Dooley Mountain rhyolite.
21.4	Perlite and flow-banded rhyolite of the Dooley Mountain rhyolite.
24.6	Turn left (east) at the road intersection toward Bridgeport and Durkee.
29.9	Turn left (north) again at the road intersection just north of Bridgeport on to the road that is parallel with Burnt River.
32.5	**Stop 1.** The rocks here are greenstones and greenschists of the Burnt River Schist. The mineral assemblage indicates greenschist facies, and the foliation (S1) is associated with D1 deformation.
32.6	**Stop 2.** The keratophyre exposed here is also part of the Burnt River Schist; the metamorphic fabric is not as well developed as at Stop 1.
34.3	**Stop 3.** The outcrops here are typical of pelitic phyllites in the Burnt River Schist and have a well-developed S1 foliation with a greenschist to sub-greenschist facies mineral assemblage. Search for crenulation of the S1 surface and any associated cleavage.
36.0	**Stop 4.** The limestone-marble of the Burnt River Schist is highly recrystallized and dolomitic. These podlike limestone bodies are enclosed in phyllite and keratophyre, and their mapped expression and shape may reflect D1 folding. Stretched pebble conglomerates occur with clasts of chert and volcanic fragments.
36.5	**Stop 5.** The rhythmically bedded, dolomitic and silicified limestone-marble is part of the Burnt River Schist and has beds that are 4 to 12 centimeters thick. Conodonts are recognizable, but they are marginally identifiable because of their deformation.
42.9	**Stop 6.** The quartz diorite pluton in the Burnt River Schist is deformed and metamorphosed to the same grade as the surrounding schists.
49.5	Turn south (right) toward Durkee.
55.1.	Join Interstate Highway 84 here. Across I-84 there is a large, highly deformed outcrop of the Nelson Marble. The poorly preserved conodonts indicate a possible Early Triassic age. The real question here is the lithologic affinity of the Nelson Marble. Is it part of the Burnt River Schist, or does it represent deposition on the Burnt River Schist prior to regional metamorphism that affected both units? The Nelson Marble has been mapped as a continuous unit and extends

several miles along strike. Continue south on I-84.

60.0 At this rest area on I-84 we have a view to the south of the Conner Creek fault escarpment which is tilted to the north.

60.4 **Stop 7.**

The Conner Creek fault is exposed here in the roadcut to the east on I-84. The fault is a high-angle reverse fault which juxtaposes Jurassic flysch of the Weatherby Formation against Permian-Triassic Burnt River Schist.

61.7 I-84 crosses the Burnt River.

67.7 **Stop 8.**

The Weatherby Formation Jurassic flysch contains limestone exposed in the quarry at Lime.

70.0 Take the exit (No. 385) here to Huntington. The distinctive red and green unit exposed on the west side of the exit is basal conglomerate of the Jurassic flysch.

73.0 Turn left past the U.S. Post Office at Huntington onto the Snake River Road and proceed across the Burnt River.

76.5 **Stop 9.**

Late Triassic volcanic and volcaniclastic rocks of the Huntington arc. Triassic flows and breccia, with some intercalated sediments, are exposed in roadcuts for the next 7 miles. This is the Spring Creek recreation site just to the right.

81.9 **Stop 10.**

Sedimentary breccias, conglomerates, siltstones, and sandstones of the Huntington volcanic arc contain angular volcanic clasts in a silty matrix.

89.8 **Stop 11.**

Examine the pelitic and fine-grained sandy beds of Jurassic flysch for comparison with rocks of the same formation approaching the Conner Creek fault to the north. The rocks here have subtle graded bedding in some exposures along the road. The character of these rocks changes little in the next 7 miles.

96.9 **Stop 12.**

The high-angle Conner Creek reverse fault is not clearly exposed. However, this is a good location to examine the fabrics of tectonically juxtaposed Jurassic flysch (foliation: N. 70° E. strike, 35° NW dip) on the uplifted south side and compare with the fabrics of the Burnt River Schist (foliation: N. 40° E. strike, 55° SE dip) on the north side of the fault. There are drag folds with apparent left lateral movement in the Burnt River Schist.

97.8 **Stop 13.**

This is an excellent, well-exposed outcrop of small-scale structure-fabric features in the Burnt River Schist. The phyllite here shows evidence for three deformational episodes, the latter of which may be related to development of the Conner Creek fault.

102.3 **Stop 14.**

The greenstone member of the Burnt River Schist contains relict titanaugite crystals (original alkalic affinity?) at this locality. The interlayered schists are with pelitic and siliceous rocks.

103.1 **Stop 15.**

Isoclinally folded, thinly bedded ribbon cherts are characteristic of the Elkhorn Ridge Argillite. The argillaceous interbeds are phyllitic. The sheared limbs of the folds have a left-lateral sense of motion.

103.4 **Stop 16.**

Alkalic pillowed greenstones associated with Elkhorn Ridge Argillite cherts are abundant along the road in this area. Pillow structures are preserved locally, and the greenstones have titanaugite.

105.8 Thinly bedded ribbon cherts can be viewed from here across the fence to the right (east) of the Burnt River Road. The cherts here are similar to those at Stop 15 but are much better exposed for purposes of examining transposition of the bedding by cleavage of uncertain chronology.

105.9 Road summit and intersection with Ruth Gulch Road offers a great view of the Snake River canyon. Note that there is a remarkable unconformity to the east across the Snake River in which Tertiary volcanic flows are above intensely deformed phyllites of the Elkhorn Ridge Argillite. Continue north along the Snake River to Richland.

114.9 Intersection of the Snake River Road with Oregon Highway 86 to Baker. Turn left (west) toward Baker.

123.2 **Stop 17.**

Inhomogeneous, deformed, fine- to medium-grained gabbro of the Sparta complex. There is a strong to faint foliation. The gabbro is intruded by quartz diorite and keratophyre veins and dikes.

124.5 **Stop 18.**

Fine-grained gabbro of the Sparta complex has both compositional and tectonite banding or flattening of pyroxenes and plagioclase. The gabbro is intruded by pyroxene pegmatites and small diorite veins. Coarse hornblende gabbro and pegmatite are on the east end of the outcrop.

125.2 **Stop 19.**

Layered gabbro of the Sparta complex has alternating plagioclase-rich and pyroxene-rich layers from 10 to 15 centimeters in thickness. Some slight tectonite flattening of mineral grains is in the plane of layering. Quartz, albite, and prehnite replace feldspar in some leucocratic layers.

130.8 **Stop 20.**

Biggs Spring diorite. The quarry has good exposure and the rocks are fresh. Here, fine- to medium-grained mafic hornblende-biotite diorite apparently is gradational to coarse, leucocratic quartz-bearing biotite diorite. Medium-grained schlieren and xenoliths of mafic hornblende diorite-gabbro are the base of the outcrop.

This is the end of the first day trip. Return to Baker.

Day 2: Road Log for Wallowa-Seven Devils Terrane Field Guide – Southern Wallowa Mountains

Mileage Description

0.0 Begin mileage at exit 304 to Interstate Highway 84 just east of the Eldorado Motel and truck stop on Campbell Street on the east side of Baker, Oregon.

1.3 Take the exit (turn right) on I-84 to Oregon Highway 86 and continue east on Oregon 86.

5.3 The Oregon Trail Memorial is on the left. Supposedly, one can still see tracks here and there made by the metal-rimmed wheels of the covered wagons carrying settlers to western Oregon in the mid to late 1800s.

9.0 The sign to Keating and Lower Powder River.

19.8 Crossing the Powder River on Oregon 86.

22.8 Road intersection. Turn left onto the Sparta Road leaving Oregon 86.

27.0 Starting here and for the next ½ mile, are views of the Wallowa Mountains, looking north (left). Continue on the Sparta Road.

27.5 Intersection of Sparta Road and U.S. Forest Service Road 70. Continue straight ahead (stay right) on the Sparta Road to Stop 21.

29.0 **Stop 21.**

The beginning of the Sparta complex albite granite or trondhjemite roadcut is on the left. Continue about 0.1 mile further to the "best" exposure (about 40 m x 5 m) of this highly altered and fractured rock. The rock is medium-grained, granular with subhedral to anhedral albite and quartz. Its color is pale tannish where weathered, to milky (siliceous?) or pale gray where fresh. Elsewhere, textures vary from granitic to crystalloblastic, cataclastic, mylonitic, and micrographic (Prostka, 1962). Representative analyses of the trondhjemite have been given by Phelps and Ave'Lallemant (1980) and two are reproduced from their Table 1:

Chemical Analyses
(Oxide Weight Percent)

	Sample Number	
Oxide	201A	193
SiO_2	72.93	72.82
TiO_2	0.39	0.43
Al_2O_3	13.40	13.22
FeO	3.06	2.90
MnO	0.13	0.09
MgO	0.56	0.54
CaO	3.28	3.16
Na_2O	3.83	3.68
K_2O	1.04	1.25
P_2O_5	0.06	0.06
H_2O	1.33	1.22

Return to the intersection of Sparta Road and U.S. Forest Service Road 70 where we will restart our mileage.

27.5 Road intersection. Turn right (north) on Road 70 leaving the Sparta Road. Outcrops are poor, but there is mostly Sparta complex albite granite on the right and Tertiary basalts on the left for the next several miles.

30.0 Cattle guard. Sign on right: "Lily White 5 mile, Eagle Creek 8 mile." Continue on U. S. Forest Service Road 70 toward Forshey Meadows just ahead.

30.5 Road intersection of USFS Roads 70 and 7005. Continue straight ahead on 70.

30.7 Road intersection of USFS Roads 70 with 200. Turn left on 200.

31.0 **Stop 22.**

Contact between Sparta albite granite and Clover Creek Greenstone. While not a good outcrop, this is the best of very few such exposed contacts between these units. Amongst the rubble on the south and in the poorly exposed roadcut on the north, are sheared, fractured, and highly altered metabasalt or spilite. The altered, leucocratic, granitic to granodioritic debris (similar to that of Stop 1) is visible amongst the metabasaltic rubble. The question here, of course, is the relationship between the two units intrusive or depositional? The greenstones are mostly (hydrothermally?) altered with abundant anhedral chloritized grains, green clusters of epidote and associated albite and hematitic oxides after ilmenite-magnetite. Note that locally, the greenstone is "fresher" appearing with subhedral relict plagioclase laths and subophitic to intergranular, subhedral relict pyroxenes (augite?) which appear as 3-4 millimeter diameter "dark green spots" in finer groundmass.

Turn around and return by the same road to U.S. Forest Service Road 70.

31.3 Intersection of USFS Road 70 with USFS Road 200. Turn left (north) onto USFS Road 70 and continue.

33.0 Surprise Springs on the left. Continue on winding USFS Road 70.

33.9 Intersection of USFS Road 70 with USFS Road 7015. Turn right on USFS Road 7015 near the head of Empire Creek and continue.

34.7 **Stop 23.**

Clover Creek Greenstone similar to that at Stop 22: dark green, very fine- to fine-grained, dense and hard, locally with identifiable elongate plagioclase laths and equant pyroxenes, both apparently volcanic relicts; otherwise recrystallized to albite, epidote, chlorite, ±prehnite-pumpellyite. Rocks for the next 2 miles to the east have been mapped as Clover Creek Greenstone by Prostka (1962), although lithologies are varied and include immature volcaniclastic sedimentary rocks. An important question here is whether the sedimentary rocks are equivalent to the Lower Sedimentary Series of Ross (1938).

35.1 **Stop 24.**

Park here and continue on foot for the next 0.5 mile to Stops 25 and 26. First observe the "granitic" rocks which are a bit of a puzzle. Note that the rocks are granular and gray-white in color; they contain clear glassy quartz, abundant milky to clear plagioclase, moderately abundant lithic clasts of mostly angular green chert-argillite to several centimeters in diameter, and green chloritic matrix. The rock is faintly banded. Could the rocks represent reworked granitic rocks of some early igneous complex such as the Sparta complex?

35.3 **Stop 25.**

Near intersection of USFS Road 7015 and USFS Road 80, which continues left and uphill. Just up USFS Road 80 the clastic rock with "granitic" texture is in contact with highly altered, chloritized and fractured Clover Creek Greenstone and very immature green volcaniclastic metasedimentary rocks interlayered with the spilitized metabasalts. The metaclastic rocks are conglomerates and grits with mainly subrounded to subangular, greenstone debris of greenish chert or chert-argillite clasts with highly chloritic matrix.

35.9 **Stop 26.**

The folded sedimentary rocks are part of a sequence that includes: (1) thin-bedded siliceous argillite, (2) quartz-rich siltstone, and (3) fine-grained sandstone. In addition, massive sandstone grits have green chloritic matrix and identifiable angular to subrounded feldspar, quartz, and lithic clasts of greenstone, and red to green argillite.

36.9 Sharp turn in USFS Road 7015 going down into Eagle Creek. Gray marble-limestone outcrops are visible along the opposite (north) side of Eagle Creek.

38.6 Bridge over Eagle Creek. Intersection of USFS Road 7015 and USFS Road 77.

38.65 Intersection of USFS Road 7015 with USFS Road 77. Turn right on USFS Road 77.

39.2 Roadcut in Martin Bridge Limestone.

39.4 Intersection of USFS Road 77 and USFS Road 7735. Keep to the left on USFS 77.

40.3 Intersection of USFS 77 with USFS Roads 360 (on left) and 375 (on right). Continue ahead on main USFS Road 77.

41.1 Large roadcut on left with bedded Martin Bridge Limestone.

41.5 Red Lower Sedimentary Series interbedded argillite-sandstone on left.

41.7 Intersection of USFS Road 77 and USFS Road 345. Turn around here at the intersection and return to the outcrop at mileage 41.5.

41.5 **Stop 27.**

The regularly layered (several centimeters thick) rocks are buff to green, thin siltstones, sandstones, grits and thicker (to 15 centimeters) red argillites with graded bedding indicating right-side-up downstream (Paddy Creek). Note flame structures of sand and argillite, and rip-up clasts of argillite in sandstone also pointing in the "up" stratigraphic (downstream) direction. The conglomerates and breccias are poorly sorted with sand-grit matrix and clasts of red argillite, metavolcanic rocks, and sparse granite. There are lots of slump structures and convolute bedding or "folds." The coarse-grained clastic rocks seem to cut irregularly into the red argillites suggesting turbidite-Bouma sequences.

Return on USFS Road 77 along Paddy Creek to mileage 35.2 and the large roadcut in Martin Bridge Limestone.

43.8 **Stop 28.**

Large roadcut of thin (a few mm to 1-2 cm), rythmically bedded, dark, biomicritic Martin Bridge Limestone with interlayers of equally thin pyritiferous silty lenses near the confluence of Eagle and Paddy Creeks is near the type section (Ross, 1938). Recrystallization of the micritic mud forms lenses of black sparry calcite. Soft sediment slump and truncation of bedding is common. In some locations, Martin Bridge fauna and sedimentological characteristics are strikingly similar to the Upper Triassic Dachstein Reef Limestone in the Austrian and German Alps and contrast with Wrangellian sequences (Stanley, 1979). Strong Tethyan affinities of the coelenterates are evident at some localities. Near here, a Norian, Late Triassic ichthyosaur of genus *Shastosaurus Merriam* collected in 1981 (from NW1/4 sec. 21, T. 7 S., R. 44 E.) has been identified and described (Orr, 1986). This genus was previously known only from northern California.

Return north on USFS Road 77 to intersection with USFS Road 7015 at the bridge across Eagle Creek.

44.35 Bridge over Eagle Creek. Intersection of USFS Road 7015 and USFS Road 77. Keep right (north) on USFS 77.

46.6 **Stop 29.**

This large roadcut in thick-bedded, micritic Martin Bridge Limestone is dark and organic looking where fresh. Thin, shaly limestone with pyritic lenses is also present, but all these beds fizz with HCl. Start at the west end of the outcrop to see synclinal, similar-type folds which to the east are warped by broad, anticlinal, folds. The later folds have axial planar cleavage. Farther east along the roadcut the sequence of beds is faulted. Drag along the fault plane indicates the western side is up.

47.1 Intersection of USFS Road 77 and USFS Road 7020 on left (bridge across Eagle Creek). Keep right on USFS 7020. Broadly folded, thin-bedded Martin Bridge Limestone is exposed on the right, just ahead.

47.2 Intersection USFS Road 77 and USFS Road 7745, which continues to the right (north) along East Eagle Creek. Keep to the left on USFS 77.

47.4 Bridge across East Eagle Creek.

47.5 **Stop 30.**

Lowermost Hurwal Formation or upper Martin Bridge Limestone. The outcrop consists of a massive limestone breccia or conglomerate with huge boulders and smaller clasts of micritic limestone. The boulders may be quite fossiliferous, with obvious corals and sponges. Clasts are angular to rounded and flattened by deformation to define a cleavage that locally becomes foliation in the marble. Although the limestone is fossiliferous, there are areas with large amounts of sparry calcite. If this exposure is lower Hurwal, then the conglomerate probably represents erosional debris from the underlying Martin Bridge Limestone. If it is Martin Bridge Limestone, then the conglomerate is simply intraformational. Continue on USFS Road 77.

48.5 Bradley Creek Road on right. Keep to the left on USFS Road 77.

48.9 O'Brien Creek Road on right. Keep to the left on USFS Road 77.

49.3 Crossing O'Brien Creek.

49.8 Crossing Bennett Creek.

49.9 Outcrops of the Hurwal Formation on the right.

50.6 More outcrops of Hurwal Formation shales and siltstones. Crossing Dixie Creek.

50.7 View of Excelsior Gulch across Eagle Creek and good outcops of the Excelsior Gulch conglomerate of the Hurwal Formation.

51.6 **Stop 31.**

The Hurwal Formation contains bedded siltstone and mudstone or fine-grained sandstone and silty argillite in this large roadcut. The beds are several centimeters to about 20 centimeters thick. There are lots of soft sediment structures, convolute bedding, and some suggestions of flame structures. Note abundant cubes of pyrite up to 1 centimeter across. The folds at the west end of the outcrop have associated axial planar cleavage. Continue on USFS Road 77.

53.2 Entrance to Tamarack campground.

53.6 USFS Road 7750 on right. Keep left on USFS Road 77.

53.7 Bridge across Eagle Creek.

53.9 Intersection USFS Road 77 and USFS Road 7755 to Boulder Park. Turn right on USFS Road 7755 to Boulder Park.

54.7 Turn left 0.1 mile West Eagle Meadows Road to outcrop of gabbro.

54.8 **Stop 32.**

The gabbro is gray, where fresh, and massive to crudely layered, containing coarse, subhedral clinopyroxene in a groundmass of medium-grained plagioclase plus clinopyroxene (30 to 40 percent). Layers several centimeters to one-half meter thick are the result of plagioclase and clinopyroxene segregation. Locally, gabbro intrudes Hurwal Formation argillites, which are modified by contact metamorphic and associated dynamic effects to schists and gneisses. The gabbro is intruded by the Wallowa batholith. Return to USFS Road 7755 and continue to Boulder Park.

54.9 Boulder Creek Road. Continue north on USFS Road 7755.

55.3 Two Color campground.

56.6 Two Color work center.

57.1 Horse corral.

58.7 **Stop 33.**

The Boulder Park landslide occurred in the spring at least five years ago. Melting snow saturated the ground and fractures in the weathered rock above Eagle Creek. The slide damaged Boulder Park Resort and dammed Eagle Creek in the resort area. The boulders brought down by the slide are leucocratic, coarse-grained tonalite with dominant plagioclase, probably 20 percent quartz, biotite, and hornblende which in places show crude alignment suggesting foliation in the margin of the batholith. The rock has a variable alkali feldspar content and grades to granodiorite.

End of field trip. Return on USFS Road 7755 to intersection with USFS Road 77.

63.3 Intersection of USFS Roads 77, 67, and 7755. Turn right on USFS Road 67, which is a continuation of USFS Road 77. Stay on USFS Road 67.

63.4 Cross Eagle Creek.

64.4 Cross Glendenning Creek.

70.4 Intersection of USFS Road 67 and USFS Road 7055 to Balm Creek Reservoir. Turn right on USFS Road 67 along Conundrum Creek.

76.6 Bridge across Big Creek.

76.7 Intersection of USFS Road 67 and the road to Mountain Creek Reservoir. Stay on USFS Road 67; continue straight ahead.

78.3 Medical Springs and intersection of USFS Road 67 with Oregon 203. Turn left (south) on Oregon 203 and continue to the interchange with I-84. Turn left on I-84 (south) to Baker.

ACKNOWLEDGEMENTS

Work leading to the results reported here by Bishop was partly funded by Geological Society of America Penrose grants 2271-77 and 2887-81, Bowes Mining Company, Sigma Xi (82-256), and the Oregon Department of Geology and Mineral Industries. We wish to thank Roger Stewart and Bill Bonnichsen for their much needed help in editing the field guide. Remaining grammatical atrocities, misspellings, and other errors are, of course, our own.

REFERENCES

Ando, C. J., W. P. Irwin, D. L. Jones, and J. B. Saleeby, 1983, The ophiolitic North Fork terrane in the Salmon River region, central Klamath Mountains, California: Geological Society of America Bulletin, v. 94, p. 236-252.

Ashley, R., 1966, Metamorphic petrology and structure of the Burnt River Canyon area, northeastern Oregon: Stanford University Ph.D. thesis, 193 p.

Ave'Lallemant, H. G., 1976, Structure of the Canyon Mountain (Oregon) ophiolilte complex and its implications for sea-floor spreading: Geological Society of America Special Paper 173, 49 p.

Ave'Lallemant, H. G., D. W. Phelps, and J. F. Sutter, 1980, ^{40}Ar-^{39}Ar ages of some pre-Tertiary plutonic and metamorphic rocks of eastern Oregon and their tectonic relationships: Geology, v. 8, p. 371-374.

Bishop, E. M., 1988, Igneous petrology of the Baker melange terrane and the Canyon Mountain complex, eastern Oregon: Geological Society of America Abstracts with Programs, v. 20, p. 407.

Blome, C. D., D. L. Jones, B. L. Murchey, and M. Lienecki, 1986, Geologic implications of radiolarian-bearing Paleozoic and Mesozoic rocks from the Blue Mountains province, eastern Oregon, *in* T. L. Vallier and H. C. Brooks, editors., Geology of the Blue Mountains Region of Oregon, Idaho, and Washington: Biostratigraphy and Paleontology: U. S. Geological Survey Professional Paper 1435, p. 79-93.

Brooks, H. C., 1979, Plate tectonics and the geologic history of the Blue Mountains: Oregon Geology, v. 41, p 71-80.

Brooks, H. C., J. R. McIntyre, and G. W. Walker, 1976, Geology of the Oregon part of the Baker 1 by 2 degree quadrangle: Oregon Department of Geology and Mineral Industries, Geological Map Series, GMS-7.

Brooks, H. C., and T. L. Vallier, 1978, Mesozoic rocks and tectonic evolution of eastern Oregon and western Idaho, *in* D. G. Howell and K. A. McDougal, editors. Mesozoic Paleogeography of the Western United States: Society of Economic Paleontologists and Mineralogists, Pacific Coast Paleogeographic Symposium 2, p. 133-145.

Brooks, H. C., M. L. Ferns, G. R. Wheeler, and D. G. Avery, 1983, Geology and gold deposits map of the northeast quarter of the Bates quadrangle, Baker and Grant counties, Oregon: Oregon Department of Geology and Mineral Industries, Geological Map Series, GMS-29.

Brown, E. H., 1977, The crossite content of Ca amphibole as a guide to the pressure of metamorphism: Journal of Petrology, v. 18, p. 53-72.

Brown, C. E., and T. P. Thayer, 1966, Geologic map of the Canyon City quadrangle, northeastern Oregon: U. S. Geological Survey Miscellaneous Geologic Investigations Map I-447.

Coney, P. J., D. L. Jones, and J. W. H. Monger, 1980, Cordilleran suspect terranes: Nature, v. 30, p. 606-611.

Dickinson, W. R., 1974. Plate tectonics and sedimentation, *in* W. R. Dickinson, editor, Tectonics and Sedimentation: Society of Economic Paleontologists and Mineralogists, Special Publication 22, p. 1-27.

———, 1979, Mesozoic forearc basin in central Oregon: Geology, v. 7, p. 166-170.

Ernst, W. G., 1968, Amphiboles: Berlin, Springer-Verlag, 278 p.

Ferns, M. L., H. C. Brooks, and D. G. Avery, 1983, Geology and gold deposits map of of the Greenhorn quadrangle, Baker and Grant counties, Oregon: Oregon Department of Geology and Mineral Industries, Geological Map Series GMS-28.

Follo, M. J., 1986, Sedimentology of the Wallowa terrane, northeastern Oregon: Harvard University Ph.D. thesis, 292 p.

Gerlach, D., 1980, Petrology and geochemistry of plagiogranite and related basic rocks of the Canyon Mountain ophiolite, Oregon: Rice University M.S. thesis, 198 p.

Gerlach, D., H. G. Ave'Lallemant, and W. P. Leeman, 1981, An island arc origin for the Canyon Mountain complex, eastern Oregon, USA: Earth and Planetary Science Letters, v. 53, p.255-265.

Gerlach, D., W. P. Leeman, E. M. Bishop, and L. Pearcy, 1988, Isotopic and trace element constraints on the timing and nature of magmatism in the Canyon Mountain ophiolite, Oregon: Geological Society of America Abstracts with Programs, v. 20, p. A158.

Gilluly, James., 1937, Geology and mineral resources of the Baker quadrangle: U.S. Geological Survey Bulletin 879, 119 p.

Himmelberg, G. H., and R. A. Loney, 1980, Petrology of ultramafic and gabbroic rocks of the Canyon Mountain ophiolite, Oregon: American Journal of Science, v. 280A, p. 232-268.

Hotz, P. E., M. A. Lanphere, and D. A. Swanson, 1977, Triassic blueschist from northern California and north-central Oregon: Geology, v. 5, p. 659-663.

Hsu, K. J., 1968, The principles of melanges and their bearing on the Franciscan-Knoxville paradox: Geological Society of America Bulletin, v. 79, p. 1063-1074.

Irwin, W. P., 1972, Terranes of the western Paleozoic and Triassic belt in the southern Klamath Mountains, California: U.S. Geological Survey Professional Paper 800C, p. 103-111.

Jones, D., N. L. Silberling, and J. W. Hillhouse, 1977, Wrangellia – a displaced terrane in northwestern North America: Canadian Journal of Earth Science, v. 14, p. 2565-2577.

Kays, M. A., M. L. Ferns, and H. C. Brooks, 1988, Metamorphism of Triassic-Paleozoic belt rocks: A guide to field and petrologic relations in the oceanic melange, Klamath and Blue Mountains, California and Oregon, *in* W. G. Ernst, editor, Metamorphism and Crustal Evolution of the Western United States: New York, Prentice Hall, p. 1098-1142.

Liou, J. C., C. O. Ho, and T. P. Ven, 1975, Petrology of some glaucophane schists and related rocks from Taiwan: Journal of Petrology, v. 16, p. 80-109.

Morris, E. M, and B. R. Wardlaw, 1986, Conodont ages for limestones of eastern Oregon and their implications for pre-Tertiary melange terranes, *in* T. L. Vallier and H. C. Brooks, editors, Geology of the Blue Mountains Region of Oregon, Idaho, and Washington: U.S. Geological Survey Professional Paper 1435, p. 59-63.

Mullen, E. D., 1978, Geology of the Greenhorn Mountains, northeastern Oregon: Oregon State University M.S. thesis, 372 p.

———, 1983, Petrology and regional setting of peridotite and gabbro of the Canyon Mountain complex, northeast Oregon: Oregon State University Ph.D. thesis, 277 p.

———, 1985, Petrologic character of Permian and Triassic greenstones from the melange terrane of eastern Oregon and their implications for terrane origin: Geology, v. 13, p. 131-134.

Nolf, B. O., 1966, Structure and stratigraphy of part of the northern Wallowa Mountains, Oregon: Princeton University Ph.D. thesis, 135 p.

Orr, W.N., 1986, A Norian (Late Triassic) ichthyosaur from the Martin Bridge Limestone, Wallowa Mountains, Oregon, *in* T. L. Vallier and H. C. Brooks, editors, Geology of the Blue Mountains Region of Oregon, Idaho, and Washington: U. S. Geological Survey Professional Paper 1435, p. 41-52.

Pessagno, E. A., Jr., and C. D. Blome, 1986, Faunal affinities and tectonogenesis of Mesozoic rocks in the Blue Mountains province of eastern Oregon and western Idaho, *in* T. L. Vallier and H. C. Brooks, editors, Geology of the Blue Mountains region of Oregon, Idaho, and Washington: U. S. Geological Survey Professional Paper 1435, p. 65-78.

Phelps, D., and H. G. Ave'Lallement, 1980, The Sparta ophiolite complex, northeast Oregon: a plutonic equivalent to low-K20 island arc volcanism: American Journal of Science, v. 280A, p. 345-358.

Prostka, H. J., 1962, Geology of the Sparta quadrangle, Oregon: Oregon Department of Geology and Mineral Industries Map, Geological Map Series, GMS-1.

Ross, C. P., 1938, The geology of part of the Wallowa Mountains: Oregon Department of Geology and Mineral Industries Bulletin 3, 74 p.

Saleeby, J. B., 1983, Accretionary tectonics of the North American Cordillera: Annual Reviews of Earth and Planetary Science, v. 15, p. 45-73.

Sarewitz, D., 1983, Seven Devils terrane: Is it really a piece of Wrangellia: Geology, V. 11, p. 634-637.

Silberling, N. J., D. L. Jones, M. C. Blake, Jr., and D. G. Howell, 1984, Tectonic terrane map of the northern Cordillera: U.S. Geological Survey Open File Report 84-523, 43 p.

Silberling, N. J., and E. T. Tozer, 1968, Biostratigraphic classification of the marine Triassic in North America: Geological Society of America Special Paper 110, 63 p.

Smith, J. P., 1912, The occurrence of coral reefs in the Triassic of North America: American Journal of Science, v. 33, p. 92-96.

Smith, J. P., 1927, Upper Triassic marine invertebrate faunas of North America: U.S. Geological Survey Professional Paper 141, 262 p.

Smith, W. D., and J. E. Allen, 1941, Geology and physiography of the northern Wallowa Mountains, Oregon: Oregon Department of Geology and Mineral Industries Bulletin 12, 64 p.

Stanley, G. D., 1979, Paleoecology, structure, and distribution of Triassic coral buildups in North America: Comparison with the Alpine Triassic of Europe: Rivista Italiana di Paleontologia e Stratigraphia, v. 85, p. 877-894.

Stanley, G.D., and B. Senowbari-Daryan, 1986, Upper Triassic, Dachstein-type reef limestone from the Wallowa Mountains, Oregon: Report on the first occurences in the United States: Palaios, v. 1, p. 172-177.

Taubeneck, W. H., 1964, Cornucopia stock, Wallowa Mountains, northeastern Oregon: Field relationships: Geological Society of America Bulletin, v. 75, p. 1093-1116.

———, 1987, The Wallowa Mountains, northeast Oregon: Geological Society of America Centennial Field Guide, Cordilleran Section, p. 327-332.

Thayer, T. P., 1977, The Canyon Mountain complex, Oregon, and some problems of ophiolites, *in* R. G. Coleman and W. P. Irwin, editors, North American Ophiolites: Oregon Department of Geology and Mineral Industries Bulletin 95, p. 93-105.

Vallier, T. L., 1977, The Permian and Triassic Seven Devils Group, western Idaho and northeastern Oregon: U. S. Geological Survey Bulletin 1437, 58 p.

Vallier, T. L., H. C. Brooks, and T. P. Thayer, 1977, Paleozoic rocks in eastern Oregon and western Idaho, *in* J. H. Stewart, C. Stevens, and A. E. Fritsche, editors, Pacific Coast Paleogeoraphy Symposium, v. 1: Society of Economic Paleontologists and Mineralogists, p. 455-466.

Vallier, T. L., and R. Batiza, 1978, Petrogenesis of spilite and keratophyre from a Permian and Triassic volcanic arc terrane, eastern Oregon and western Idaho, U.S.A.: Canadian Journal of Earth Science, v. 15, p. 1356-1369.

Walker, G., 1979, Reconnaissance geologic map of the Oregon part of the Grangeville quadrangle, Baker, Union, Umatilla, and Wallowa Counties, Oregon: U. S. Geological Survey Miscellaneous Investigations Map I-902.

Walker, N. M., 1983, Pre-Tertiary tectonic evolution of northeastern Oregon and west-central Idaho: Constraints based on U/Pb ages of zircons: Geological Society of America Abstracts with Programs, v. 15, p. 371.

Chapter Three
Tectonic and Sedimentary Sequences in Northeast Idaho and Northwest Montana

Moyie River Falls near Bonners Ferry, Idaho, cuts through rocks of the Precambrian Belt Supergroup. *Photograph courtesy of Idaho Division of Travel Promotion.*

A Structural Section Through a 25-Km-Thick Thrust Plate in West-Central Montana: A Field Trip From Paradise to Garrison

James W. Sears[1]
Christopher P. Weiss[1]
Patrick H. Reynolds[1]
Joseph H. Griffin[2]

INTRODUCTION

This field trip examines the changes in structural style that occur with changing depth in the western Montana thrust and fold system. The system plunges continuously to the southeast on the limb of a large south-facing monocline (Figure 1). The trip starts at the structurally deep northwestern end of the plunging system near Paradise and finishes at the shallow southeastern end near Garrison (Figure 2). In the northwest, the system is in the biotite zone of regional metamorphism. In the southeast it is a foreland basin with syntectonic conglomerates derived from the thrust system. In between it is a complete transition from the ductile interior of the thrust system to the brittle carapace. Parts of this trip were previously included in a Tobacco Root Geological Society field trip (Sears, 1988a).

[1]*Department of Geology, University of Montana, Missoula, Montana 59812.*
[2]*Hunter/ESE, P.O. Box 1703, Gainesville, Florida 32602-1703.*

TECTONIC SETTING

This field trip follows the structurally complex overlap zone between two major thrust slabs (Sears, 1988b). The western slab forms the Sapphire and Bitterroot Mountains; the eastern slab forms the mountains from Missoula east to the Rocky Mountain front. The western slab includes the Sapphire tectonic block of Hyndman (1980) and Hyndman and others (1988) and the Rock Creek

Sears, J. W., C. P. Weiss, P. H. Reynolds, and J. H. Griffin, A structural section through a 25-km-thick thrust plate in west-central Montana: A field trip from Paradise to Garrison, *in* V. E. Chamberlain, R. M. Breckenridge, and Bill Bonnichsen, editors, Guidebook to the Geology of Northern and Western Idaho and Surrounding Area: Idaho Geological Survey Bulletin 28, p. 87-102.

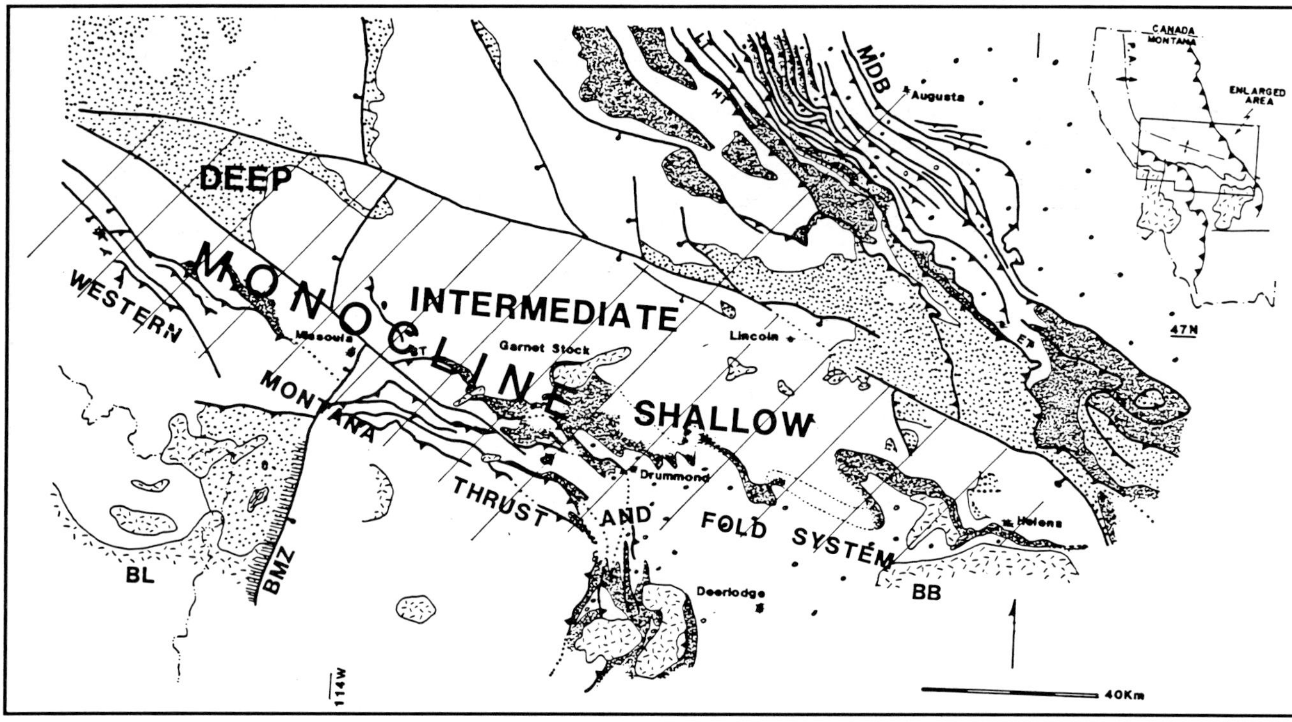

Figure 1. Location map for field trip. Symbols: Open stipple—lower part of Belt Supergroup; unpatterned—upper part of Belt Supergroup; coarse stipple—Paleozoic section; circles—Mesozoic rocks; chicken tracks—Late Cretaceous plutons; horizontal lines—Bitterroot mylonite zone; wide cross-hatching—monocline. Abbreviations: BB—Boulder batholith; BL—Bitterroot batholith; BMZ—Bitterroot mylonite zone; MDB—Montana disturbed belt; PA—Purcell anticlinorium. After Sears, 1988a.

subplate of Ruppel and others (1981).

The western slab was emplaced during the Upper Cretaceous Campanian Age, about 82 million years ago, along a zone of cleavage, folds, and imbricated thrusts (Figure 3a). Important thrust faults carrying the western slab are the Lothrop and Albert Creek thrusts, west of Missoula (Hall, 1968; Lonn, 1986), the Bearmouth and Blackfoot thrusts, east of Missoula (Nelson and Dobell, 1961; Kauffman, 1961), and the Philipsburg and Georgetown thrusts, south of Drummond (Emmons and Calkins, 1913).

The eastern slab overrode the Montana disturbed belt along the Lewis, Hoadley, and Eldorado thrusts by late Paleocene time (Mudge and others, 1982) and carried the western slab piggyback (Figure 3b). This post-Campanian thrust movement rotated the earlier-formed structures of the overlap zone and produced systematic regional plunge toward the southeast, possibly where the slabs drape a major transverse footwall ramp (Sears, 188b). Pre-middle Eocene erosion bevelled and exposed the plunging thrust system. Middle Eocene volcanic rocks, dikes, and high-level intrusions overlap or crosscut the structures (Hyndman and others, 1988). Post-middle Eocene block faults along the Lewis and Clark line disrupted the thrust system and rotated the Late Cretaceous deformational fabrics (Figure 3c).

Table 1 summarizes the stratigraphy of the field trip area. The Middle Proterozoic Belt Supergroup forms the bulk of both thrust slabs, with a thickness estimated to be nearly 20 kilometers in the eastern plate near Missoula (Harrison, 1972). Paleozoic shelf strata up to 2 kilometers thick and a Mesozoic foreland basin succession up to 5 kilometers thick cap the Belt Supergroup in the western Montana thrust and fold system (Kauffman, 1961; Gwinn, 1961; Gwinn and Mutch, 1965).

The invasion of granitic magma changed regional geothermal conditions in west-central Montana during the emplacement of the western slab (Hyndman and others, 1988). Cool, brittle rocks initially formed the leading edge of the western slab along the Albert Creek, Bearmouth, Philipsburg, and associated thrusts. The Missoula Group and Wallace Formation of the Belt Supergroup overrode Paleozoic and Mesozoic strata and produced breccias and gouge zones. The uplifted western slab shed coarse clasts into the Campanian Golden Spike Formation in the adjacent foreland basin by about 82 Ma (Gwinn and Mutch, 1965; Ruppel and others, 1981; Mackie, 1986). Unstrained clasts from as deep as the Bonner Formation accumulated in the "chaos beds," a diamictite deposit within the Golden Spike Formation near Garrison.

As the western slab continued to move, the footwall

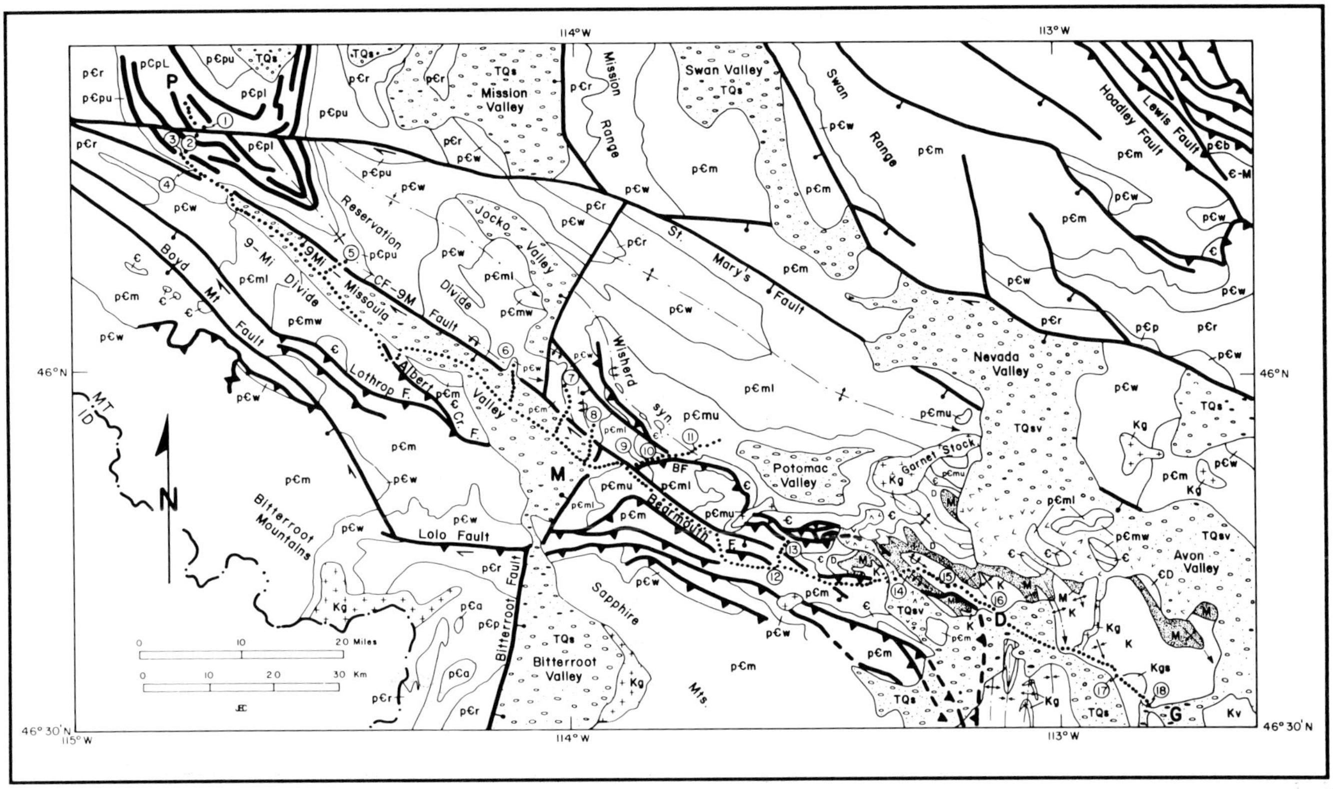

Figure 2. Generalized geologic map of field trip area. Coarse dotted line shows route of field trip. Figure 1 gives location. Symbols: pЄa, pre-Belt crystalline rocks?; pЄb, Belt Supergroup; pЄpl, lower Prichard Formation; pЄpu, upper Prichard Formation, pЄr, Ravalli Group; pЄw, Wallace and Helena Formations; pЄml, lower Missoula Group; pЄmu, upper Missoula Group; pЄm, Missoula Group; Є, Cambrian; D, Devonian; M, Mississippian-Permian; K, Jurassic-Cretaceous; Kgs, Golden Spike Formation; Kg, Late Cretaceous plutons; TQ, Eocene volcanics and Tertiary and Quaternary sediments. Black bands are diabase sills. After Harrison and others, 1986; Mudge and others, 1982; Nelson and Dobell, 1961; Hall, 1968; Lonn, 1986; Thomas, 1987; Desormier, 1975; Gwinn, 1961; Kauffman, 1961; Maxwell, 1965; Myers, 1986; Nold, 1968; Wallace and others, 1981; Watson, 1984. Units are not differentiated in Montana disturbed belt in northeastern part of map.

rocks became more plastic. A greenschist facies mylonite zone and widespread cleavage formed at stratigraphic levels which had initially been brittle. The cleavage formed during and after growth of cordierite porphyroblasts in the contact aureole of the Garnet stock (Minnich, 1984), which has a hornblende K-Ar age of 82 Ma (Ruppel and others, 1981). The cleavage affects the matrix of the Golden Spike diamictites.

We have traced the cleavage zone from the Missoula area eastward to the Big Belt Mountains and southward to the Flint Creek Range. Similar Campanian-aged cleavage occurs in the Pioneer Mountains, in the footwall, and east of the Grasshopper thrust plate (Sears and others, 1988). The cleavage is strongest adjacent to the border of the western plate and near granitic bodies, where it is locally a biotite schistosity.

The route of this field trip crosses a major boundary in Late Cretaceous geothermal conditions (Weiss, 1987). The crust between Garrison and Missoula had a gradient of at least 40° or 50° C per kilometer of depth. The crust from Missoula to Paradise had a lower average geothermal gradient, because the rocks at an estimated depth of 25 kilometers near Paradise are in the biotite-chlorite zone of regional metamorphism, between 450° and 500° C (Hyndman, 1986). This gradient is slightly above Turcotte and Schubert's (1982) average continental geotherm, which indicates rocks at a depth of 25 kilometers are typically at 350° C. Extrapolation of the geothermal gradient from Garrison to Missoula to greater depth would require the Prichard Formation near Paradise to be in the upper amphibolite facies. Prichard rocks south of the thermal boundary near the Bitterroot batholith are in the sillimanite-migmatite zone.

The region of high temperature-low pressure Late Cretaceous metamorphism envelops abundant coeval igneous rocks and indicates a genetic link among igneous activity, metamorphism, and growth of the associated structures. Although the oldest sills share all deformation, the major plutons crystallized after the fold-thrust structures formed.

Figure 4 is a composite section through the area of the field trip. It stacks cross-sections from different levels of

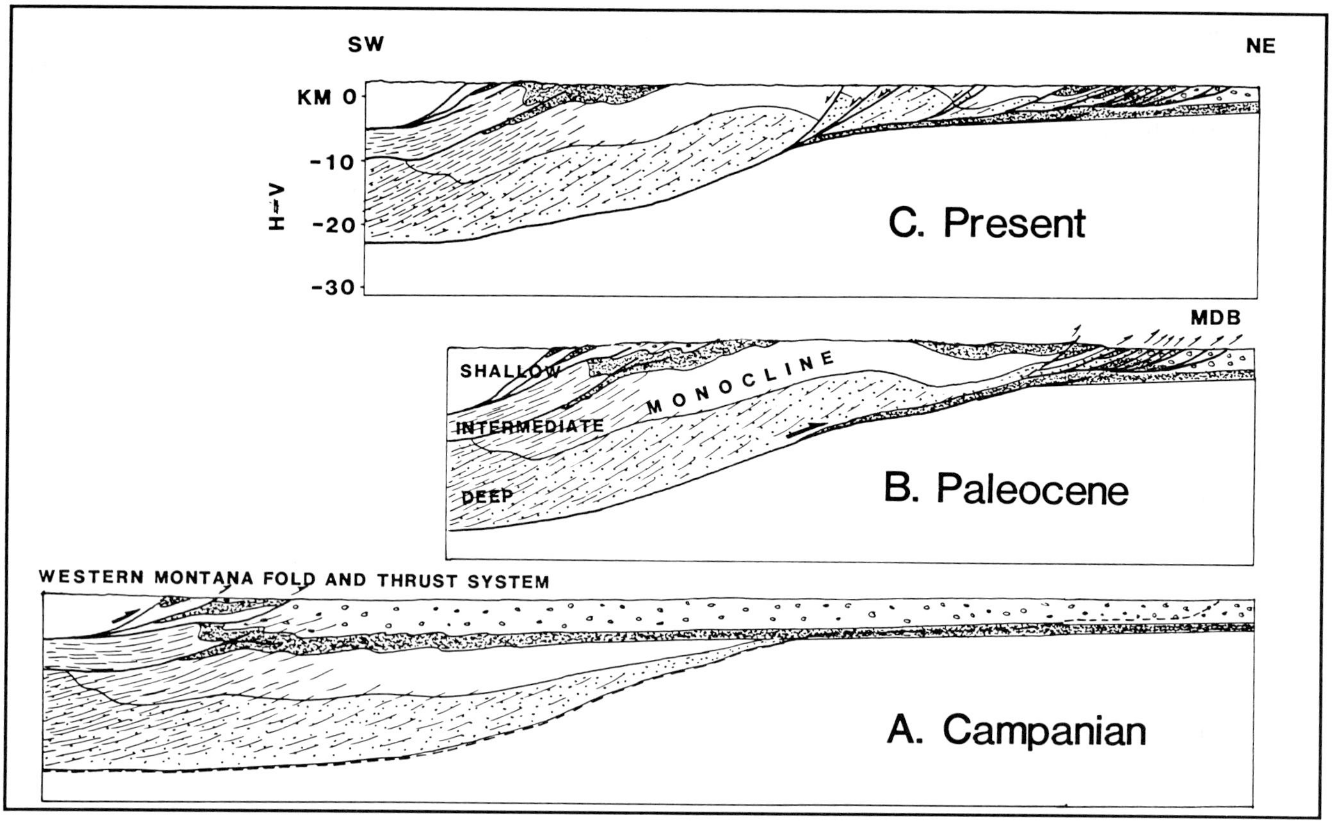

Figure 3. Cross-sections along line passing through Augusta (see Figure 1), showing the development of western Montana fold and thrust system. Same symbols as Figure 1. After Sears, 1988a.

Table 1. Stratigraphic units in field trip area. Symbols are the same as those used on Figure 2 and Figure 4. Data taken from Cressman, 1985; Gwinn, 1961; Kauffman, 1961; Watson, 1984; Weiss, 1987.

Symbol	Thickness in km	Rock Unit
TQsv		Tertiary and Quaternary poorly consolidated basin fill and Middle Eocene volcanic rock
		Mesozoic
Kgs	1.9	Golden Spike Formation
Ku	3.5	Blackleaf, Coberly, Jens, Carter Creek Fms
JKu	0.5	Ellis Group and Kootenai Formation
	5.9	
		Paleozoic
PPu	0.2	Amsden, Quadrant and Phosphoria Fms
Mm	0.7	Madison Group
Du	0.7	Devonian rocks
Єu	0.8	Flathead, Silver Hill, Hasmark, and Red Lion
	2.4	

Symbol	Thickness in km	Rock Unit
		Precambrian Belt Supergroup
		Missoula Group (pЄm)
pЄpi	0.4	Pilcher Quartzite
pЄgr	1.2	Garnet Range Formation
pЄmc	1.2	McNamara Formation
pЄbo	0.4	Bonner Quartzite
pЄms	1.6	Mount Shields Formation
pЄsh	0.5	Shepard Formation
pЄsn	1.1	Snowslip Formation
	6.4	
		Middle Belt Carbonate
pЄw	2.7	Wallace and Helena Formations
	2.7	
		Ravalli Group (pЄr)
pЄe	0.4	Empire Formation
pЄsr	0.6	St. Regis Formation
pЄre	0.9	Revett Formation
pЄbu	1.0	Burke Formation
	2.9	
		Lower Belt
pЄpu	1.6	upper Prichard Formation
pЄpl	4.4	lower Prichard Formation
	6.0	
Grand total	**26.3**	

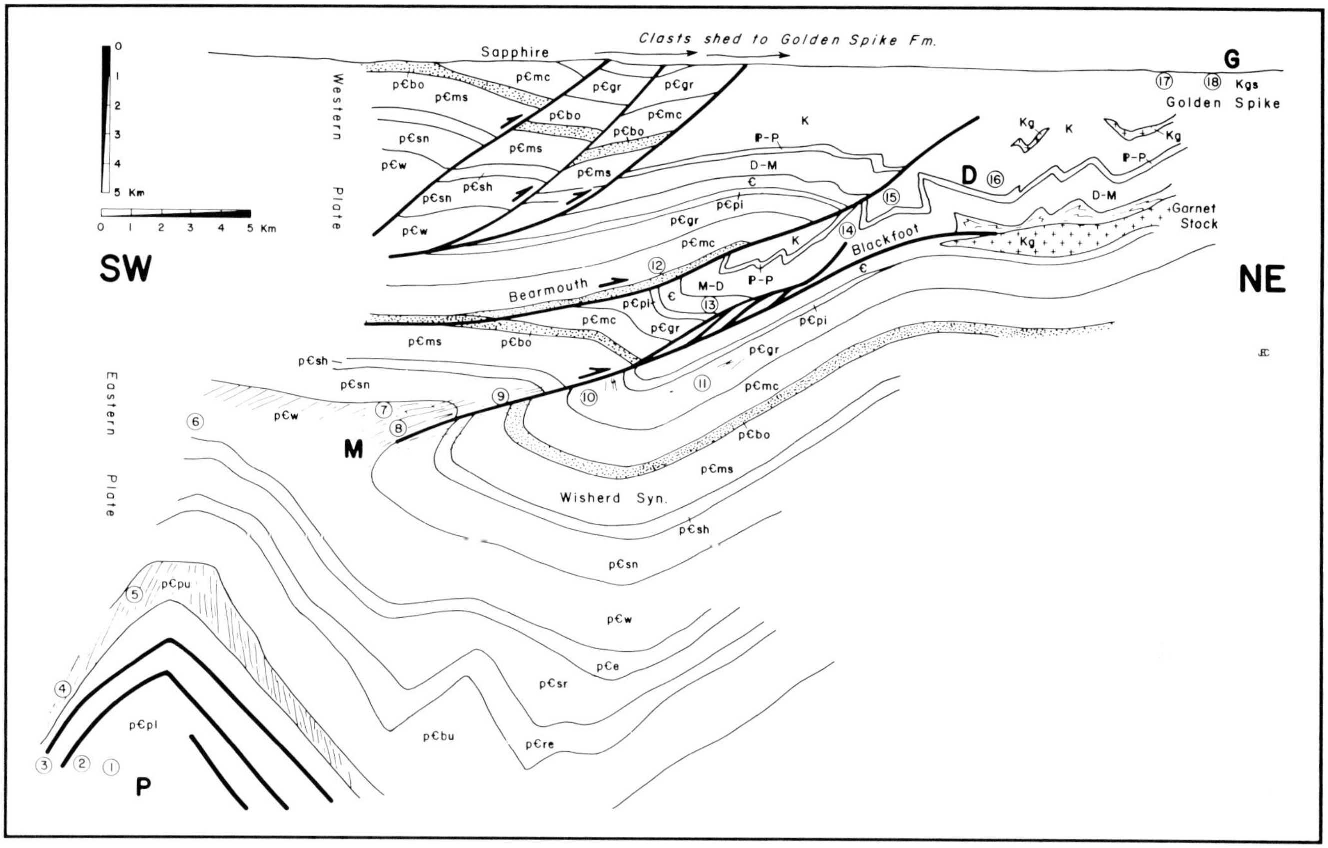

Figure 4. Composite down-plunge projection of the field trip area, showing approximate projected locations of field trip stops. No vertical exaggeration intended. See Table 1 for stratigraphic symbols. P, Paradise; M, Missoula; D, Drummond; G, Garrison.

the thrust system along the regional plunge. Approximate locations of field trip stops are shown where they project on the composite section.

The composite section shows three distinct levels of deformation in the system. The lowest level is a large kink fold of the siliceous lower Belt Supergroup. The middle level contains a nappe of the middle Belt carbonate, detached from the deeper kink folds. The upper level is a thrust-fold system in the upper Belt Supergroup and Phanerozoic section, which in part grew from the middle level nappe. The behavior of each level depended on the dominant rock type and on the geothermal conditions at the time of deformation.

ROAD LOG

Day 1: Paradise to Bonner

Mileage Description

0.0 The roadlog starts at the U.S. Post Office in Paradise. The log crosses Seigel Pass on secondary U.S. Forest Service Road 412. Enquire locally about road conditions. If the pass is closed, use an alternate route between Stops 3 and 5 (Figure 5).

Paradise is in the core of a large irregular dome, a culmination of the Purcell anticlinorium. Price (1981) and Bally (1984) showed the Purcell anticlinorium to be a major thrust-ramp anticline, formed where thick basinal deposits overthrust the cratonal margin. The Paradise dome is at a dogleg bend in the Purcell anticlinorium (Figure 1, inset). The field trip follows folds which plunge southeast, down the monoclinal back limb of the anticlinorium.

Tertiary right-slip faults of the Lewis and Clark line segment the dome. The St. Maries fault offsets the southern part relatively westward near Paradise (Harrison and others, 1986).

The dome exposes the Prichard Formation, the lowest part of the Belt Supergroup. Cressman (1985) reported the Prichard Formation to be about 7 kilometers thick in this area, including several thick mafic sills. The base of the

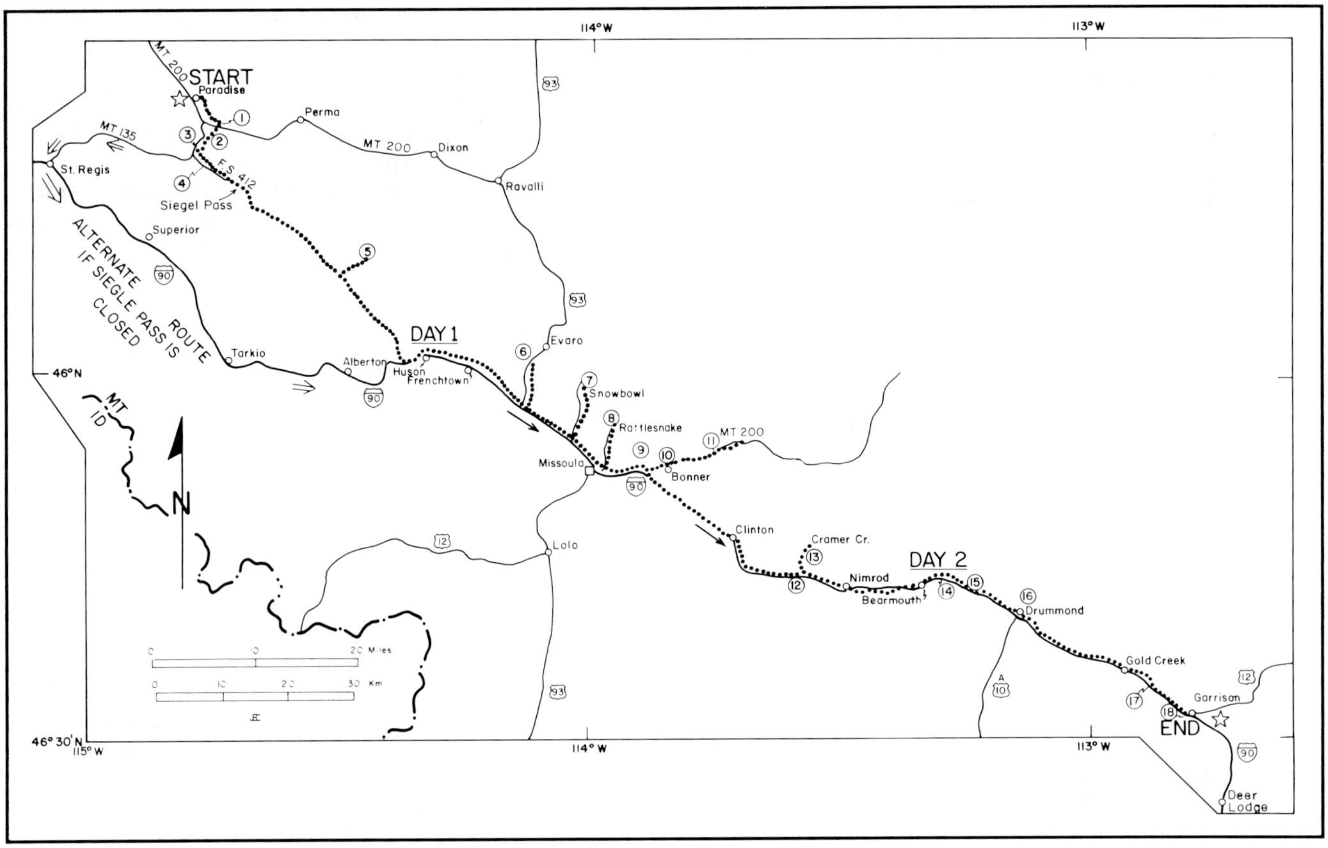

Figure 5. Map of field trip route. Route is shown by dotted line. Stops are shown by circled numbers.

Prichard Formation is not exposed in this area.
Proceed southeast on Montana Highway 200.

0.5 Bridge across the Clark Fork River.

2.7 Bridge across the Clark Fork River. The prominantly jointed cliff ahead and to the left is a thick diabase sill within the Prichard Formation.

3.0 Turn right at junction on Montana Highway 135. Proceed south.

3.3 **Stop 1: Prichard Formation in regional biotite metamorphic zone.**

Stop at the first roadcut on the left. A one-lane road is on the right, across from the roadcut.

This exposure is the basal part of the Prichard Formation, member A of Cressman (1985), the deepest exposed part of the thrust system. From here, the system plunges steadily to the southeast. Stacking the section along the plunge suggests about a 20-kilometer structural thickness for the Belt Supergroup between here and Missoula (Figure 4).

This stop is in the core of a large, southeast-plunging, kink-fold anticline. The anticline has a steeply dipping axial plane at this level; but as shown in Figure 4, at higher levels the axial plane dips gently in the softer rocks of the middle Belt carbonate, and the fold is a nappe.

The Prichard Formation is in the biotite-oligoclase zone of burial metamorphism in most of western Montana, with local areas in the garnet-amphibole zone (Norwick, 1972). Burial metamorphic minerals typically follow bedding or are irregularly oriented and cross undisturbed bedding laminations. Burial metamorphic biotites have K-Ar ages of 1.3 Ga (Obradovich and Peterman, 1968).

At this outcrop, fine-grained biotite forms a schistosity at a high angle to the bedding in pelitic layers. This schistosity dips southwest and conforms to the orientation of map-scale folds between here and Missoula. We have traced this cleavage surface from here to Drummond, where it cuts Cretaceous rocks. We interpret this to be Late Cretaceous regional biotite zone metamorphism.

Proceed southwest on Montana 135.

4.1 Enter Lolo National Forest. The gorge of the Clark Fork River exposes a thick section of the

Prichard Formation. For the next 3 miles the highway crosses generally up section through the Prichard Formation.

4.6 Note the overturned, southeast-plunging syncline in the roadcut on the left. Some folds in this area are syndepositional (Cressman, 1985).

5.2 Quinn Hot Springs is on the left.

5.4 **Stop 2: Polyphase metamorphism of the Prichard Formation.**

Stop at the first roadcut on the left, south of Quinn Hot Springs. Note survey marker witness post and U.S. Geological Survey benchmark D520 at the outcrop.

This stop shows the relationship between two periods of metamorphism in the Prichard Formation. The bedding dips steeply to the southwest on the southwest limb of the major kink-fold anticline. The quartzite beds preserve delicate depositional and diagenetic structures and are uncleaved. The layers of mica schist are completely recrystallized. Porphyroblasts of biotite in the schist are retrograded to chlorite, deformed, and rotated into the cleavage, which is defined by aligned muscovite grains. Here the cleavage dips to the southwest more steeply than the bedding, and it is systematic with the axial plane of the map-scale kink fold.

We interpret the porphyroblasts to be altered burial metamorphic biotites. The schistosity formed during the Late Cretaceous. We have passed out of the Late Cretaceous biotite zone of Stop 1 and into the Late Cretaceous muscovite-chlorite zone.

Continue southwest on Montana 135. Proceed steadily up-section through the Prichard Formation on the southwest limb of the large kink fold.

6.2 Camp Bighorn is on the left.

6.8 Milepost 18. Proceed around curve to the left.

7.3 **Stop 3: Deformed mafic sill in the Prichard Formation.**

Stop at large pullout on the right, overlooking Clark Fork River.

The thick mafic sill exposed here is on the southwest limb of the large kink fold. Bedding in the enclosing Prichard Formation dips steeply to the southwest and is visible both to the left and right of the sill. The sill exhibits quartz- and calcite-filled tension gashes in an en echelon pattern consistent with shear on the limb of the anticline (top to the northeast). Other parts of the sill are brecciated, with vein material between fragments. There are local zones of chlorite schist with southwest-dipping schistosity and weak southwest-plunging lineation.

We interpret the deformation of this sill to be Late Cretaceous, because it is consistent with the regional folding. The mineral assemblage suggests that lower greenschist facies (chlorite zone) conditions existed in these rocks at the time of deformation. As at Stop 2, Prichard Formation quartzite lacks evidence of plastic deformation.

Turn around. Proceed northeast on Montana 135.

7.8 Turn right at Milepost 18 onto Forest Service Road 412 and proceed southeast along Siegel Creek.

If Siegel Pass is closed, follow Montana 135 to St. Regis, Interstate 90 to Ninemile exit, and Ninemile Road to McCormick Creek, for Stop 5. This longer way around is about 75 miles. It follows the Boyd Mountain fault strand from St. Regis to Tarkio, then crosses a thrust and fold system along the Albertan Gorge from Tarkio to Ninemile. For further information on this route, see Winston and Lonn (1988).

The Clark Fork-Ninemile fault, one of the major strands of the Lewis and Clark line, follows the south wall of Siegel Creek valley. The Forest Service road climbs to Siegel Pass and also up section through the Prichard Formation.

13.9 **Stop 4: Upper Prichard Formation.**

Stop at hairpin switchback. The upper Prichard Formation is a structurally weak phyllite that deformed more plastically than the underlying quartz-rich Prichard or overlying quartzites of the Ravalli Group. It forms a detachment zone and accommodates disharmonic folding between the lower Prichard Formation and the Ravalli Group. Note in Figure 4 that the sharp kink fold in the Prichard Formation broadens in the Ravalli Group.

Proceed southeast on Siegel Creek Road.

16.9 Siegel Pass. Continue southeast on Forest Service Road 412 toward Ninemile Ranger Station. Ninemile Valley follows the Clark Fork-Ninemile fault zone. This fault had right-lateral slip with a component of down-to-the-south normal slip during Tertiary time, and formed the Ninemile and Missoula valleys.

20.1 Junction with Forest Service Road 97. Bear left and remain on Road 412 toward Ninemile Ranger Station.

21.9 Ninemile Divide-Superior road enters from right.

22.2 Prichard Formation crops out on left. Cross Beecher Creek.

27.9 Foothills Road branches off to the left. Bear right.

28.7 Cross cattle guard.

32.7 West Side Ninemile Road on right. Proceed straight.

33.7 Cross McCormick Creek.

33.9 Turn left on Forest Service Road 392 and proceed northeast up McCormick Creek.

34.9 Tailings from placer operations line McCormick Creek.

35.8 Bear left and remain on Road 392.

36.1 Cross McCormick Creek near active gold dredging operations.

36.2 Turn right and remain on Road 392.

37.6 **Stop 5: Fabric of upper Prichard Formation.**

Park at junction of Road 392 and Road 16238, at tight curve. Walk along the old road up McCormick Creek 1/4 mile to Stop 5.

Outcrops along the old road show very steep southwest-dipping phyllitic cleavage crossing gentle southwest-dipping bedding in the upper Prichard Formation. Bedding laminations were nearly passive during growth of the cleavage. Cleavage planes have distinct stretching lineations nearly normal to bedding-cleavage intersections. Quartz fringes on pyrite grains show maximum elongations of 3 to 7 in the stretching direction (Weiss, 1987).

About 1/4 mile farther up the old road, a small outcrop on the left shows distinctly curved bedding-cleavage intersection lineations on cleavage surfaces. The intersection lineation arches in the direction of the stretching lineations.

Cross McCormick Creek on the old road and continue walking for about 1/4 mile. The first outcrop on the right shows an overturned fold of bedding with axial planar cleavage. This is a parasitic fold on the limb of the major fold we discussed at the first four stops. The axial trace of the major fold crosses McCormick Creek about 1/4 mile upsteam from here.

Return to vehicles and go back to Ninemile Road.

41.3 Junction of Road 392 and Ninemile Road. Turn left on Ninemile Road.

42.9 Ninemile Community Center is on the left.

43.8 Garnet Range Formation crops out on the left. We are south of the trace of the Clark Fork-Ninemile fault, which here places the Ravalli Group against the Garnet Range Formation, cutting out a 10 kilometer thickness of Belt Supergroup. We believe the Garnet Range Formation seen here lies on the same thrust plate as the rocks north of the Clark Fork-Ninemile fault.

46.9 West Ninemile Road enters from the right.

49.7 Ninemile Ranger Station. Stay on main road to right.

50.1 Pavement begins. Road passes up section through Pilcher Formation, which forms hill on the right.

50.6 Yellow frame buildings of the "Schoolhouse Teacherage" are on the right. Cambrian Hasmark Dolomite crops out on left. This is in the footwall of the Albert Creek thrust, which crosses the road just south of here, placing the Pilcher Formation over the Hasmark Dolomite (Wells, 1974). The Albert Creek fault is one of the faults at the leading edge of the western slab. It passes into a syncline a few miles northwest of here, perhaps transferring displacement to the structurally overlying Lothrop thrust (see Lonn, 1986).

52.3 Intersection with old U.S. Highway 10. Turn left.

52.5 Sediments of Pleistocene glacial Lake Missoula crop out.

52.8 Garnet Range Formation crops out in hanging wall of Albert Creek thrust.

53.8 Go under overpass and turn left onto I-90 east, toward Missoula. The trace of the Albert Creek thrust is near the east end of the entrance ramp, here placing Garnet Range Formation over Cambrian Red Lion Formation.

55.2 Near Milepost 84, Pilcher Formation crops out in quarried hill on the right, beside the Clark Fork River. Across the river in the distance is a prominent cliff of the Cambrian Red Lion Formation. The more distant hills are Missoula Group rocks in the hangingwall plate of the Albert Creek thrust.

56.2 Huson overpass, Exit 85. Stay on I-90. The Missoula Valley is a complex half-graben tilted toward the northeast against the Clark Fork-Ninemile fault. Middle Eocene volcanic rocks remain as patches on the hills to the right, across the river (Sears, 1985). They rest on an angular unconformity and overlap the Albert Creek thrust (Hall, 1968). They dip into the valley and pass beneath thick Tertiary valley fill.

59.5 Frenchtown Pond is on the left.

62.4 Glacial Lake Missoula sediments crop out to the left.

65.7 The roadcuts on the left along the frontage road are the best exposures of Tertiary strata in the Missoula Valley.

66.5 Stone container plant is on the right. The cliffs behind the plant are Cambrian Hasmark Dolomite. The higher elevations are Missoula Group rocks in the hanging wall of the Albert Creek thrust, which follows the break in slope. The thrust passes beneath the valley fill about 2 miles southeast of here, carrying Mount Shields Formation over Hasmark Dolomite. East of Missoula, in the Sapphire Range, the Bearmouth thrust may be the continuation of the Albert Creek thrust. It places Mount Shields Formation over Garnet Range Formation and Bonner Quartzite over Hasmark Dolomite.

66.8 Take exit 96.

67.0 Turn left on U.S. Highway 93N. Traverse glacial Lake Missoula sediments and Tertiary valley fill. Railroad cuts in hills to right expose steeply tilted Tertiary sandstones, shales, coal beds, and bentonites of the Renova Formation.

70.8 Cross trace of Clark Fork-Ninemile fault.

71.0 **Stop 6: Refracted cleavage in the Wallace Formation.**

Mouth of canyon of O'Keefe Creek at beginning of steep grade. Park at pullout beyond the chain-up area on the right. Cross U.S. 93 to large outcrop by powerpole.

This large outcrop of calcareous pelites and quartzites is on the southwest limb of the large anticline we visited on Stops 1-5. These rocks are about 6 kilometers higher in the section than those at Stop 5.

Cleavage is very strong in the calcareous and pelitic lithologies, but it is absent or widely spaced in quartzites, suggesting the rock was in the plastic field for pelite and carbonate but in the brittle field for quartzite. Cleavage refracts through lithologies of contrasting ductile strength. It nearly parallels bedding in limestone layers, lies at a shallow angle to bedding in pelites, and steepens with increasing quartz content in graded beds. Each quartzite/pelite contact is effectively a miniature ductile shear zone, with the top consistently to the northeast, because of the position of the rocks on the southwest limb of the northeast-verging anticline.

This outcrop is close to the transition between the upright, kink-like fold at depth and the nappe-like fold at the middle level of the structure. Along this transect, the anticline is steeply overturned to the northeast, and the northeast limb is cut by a thrust fault. Farther southeast, the overturning increases, and the lower limb becomes sheared out, as we shall see at Stop 8.

Return to I-90.

75.2 Turn left on I-90 and proceed east towards Missoula.

77.2 Cross Butler Creek outwash plain.

78.8 To the right glacial Lake Missoula sediment forms the terrace upon which the airport is located.

80.5 Take Grant Creek exit 101. Turn left at the stop sign and proceed under the overpass. Go northeast on Grant Creek Road.

82.5 Cross approximate trace of Clark Fork-Ninemile fault. Suburb development is on the right.

84.2 Turn left and follow signs toward Montana Snowbowl.

84.4 Outcrop of highly sheared Wallace Formation is on the right. This is on the overturned limb of the fold nappe. Bear right at junction.

84.8 Quaternary gravels crop out at tight switchback.

86.0 Road crosses pass into Butler Creek canyon.

87.4 **Stop 7: Metamorphosed Wallace Formation.**

Stop at large, light gray outcrop on the right just past the sharp curve to left.

Regional metamorphism converted the Wallace Formation into schist at this locality. Schistosity dips steeply to the southwest and crosses bedding at a high angle. Bedding is clearly visible on schistosity surfaces as brown and gray laminations and as aligned aggregates of small brown spots. The spots are probably diagenetic nodules and are deformed into crude ellipsoids. The long axes of the ellipsoids form a stretching lineation on the surface of the schistosity, approximately normal to the average bedding-cleavage intersection lineation. The bedding-cleavage lineations are warped in the schistosity.

The fabric elements at this outcrop strongly resemble those seen at Stop 5 in McCormick Creek. The rock here is also on the southwestern limb of the anticline. We are at about the same stratigraphic level as Stop 6, but clearly the grade of metamorphism is higher here. We are crossing the lateral geothermal gradient discussed in the Introduction, as we approach the metamorphic envelope of the Late Cretaceous batholith zone.

Return to I-90.

94.3 Turn left on I-90 E.

98.2 Take exit 105. Turn left at stop sign on Van Buren Street and proceed north on Rattlesnake Creek.

Van Buren Street becomes Rattlesnake Drive and follows the outwash plain of Rattlesnake Creek for the next 3 miles. Waterworks Hill on the west side of the valley exposes Snowslip and Wallace Formations; Mount Jumbo on the east side is mostly Mount Shields Formation, with Shepard Formation near the base. Right hand turn in the road at Lincoln Woods subdivision is approximately at the trace of the Ninemile-Clark Fork fault. Canyon narrows in Wallace-Helena Formation.

102.3 Turn left on Sawmill Gulch Road to trailhead of Rattlesnake National Recreation and Wilderness area.

102.6 **Stop 8: Wallace and Helena Formations.**

Park in parking area. Walk north along trail in valley bottom about 1⁄4 mile to large cliff face of Wallace-Helena Formation on the left.

Calcareous pelites of the Wallace and Helena Formations are highly sheared in this outcrop and in most outcrops along Rattlesnake Creek as well as over a large area to the northwest. These rocks occupy the lower limb of a large fold nappe that is overturned to the northeast. The bedding parallels cleavage, and there is a strong stretching lineation in the transport direction. This anticline appears to be the same one that we visited at the previous stops. The nappe apparently passes upward into the Blackfoot thrust fault, a major structure of the Garnet Range.

Return to I-90.

107.0 Turn left on I-90 E and proceed toward Butte.

107.3 I-90 crosses the Mount Sentinel Tertiary normal fault and enters Hellgate Canyon. The eastern block rose at least 5 kilometers relative to the western block, which forms the Missoula Valley. Calcareous argillites of the Shepard Formation form the brown-weathering outcrops at the west edge of the canyon. The Mount Sentinel fault abuts the Clark Fork-Ninemile fault north of here in the saddle of Mount Jumbo.

108.4 Mount Shields Formation crops out on the left. Hellgate Canyon exposes a continuous section from the upper part of the Shepard Formation through the lower part of the MacNamara Formation. These rocks form the footwall of the Bearmouth thrust. Pelitic layers locally exhibit west-dipping phyllitic schistosity.

111.1 Take exit 109 and proceed east on Montana Highway 200.

112.1 Bridge over Blackfoot River, entering Milltown. Continue on Montana 200.

112.5 Bear left and remain on Montana 200 east.

112.9 Town of Bonner. Champion plywood mill is on the left. The Blackfoot thrust lies at the base of Bonner Mountain, the forested mountain on the right. This important fault places Wallace Formation and lower Missoula Group over upper Missoula Group and Paleozoic rocks. At Bonner, the Wallace Formation on the southwest flank of Bonner Mountain (near the letter "B" on the hillside) is very strongly sheared on the overturned limb of the Bonner Mountain anticline. A good outcrop along the abandoned railway behind Bonner Elementary School shows the bedding of the Wallace Formation transposed into the shear foliation.

113.6 Montana 200 passes through a notch on a tight curve to the right. Proceed around curves.

113.8 **Stop 9: Mylonite in Blackfoot thrust fault zone.** Park on wide left shoulder of highway and walk back to the notch.

The Blackfoot thrust crops out in this long roadcut. The trace of the fault is under the highway, but because the fault dips to the southeast, small klippen occur on the north side of the road. The south side of the road is highly sheared, lower Missoula Group quartzite and argillite in the hanging wall. The north side is mostly the footwall, here a vertical diabase sill in the upper part of the Mount Shields Formation. Shearing along the fault converted the diabase into a chlorite-actinolite-epidote mylonite, with a strong mineral lineation trending northeast. Crustal temperatures were at least 350° C during thrust faulting here. The mylonitic foliation dies out within a few meters below the Blackfoot fault plane.

The vertical beds of the footwall are obvious on the north side of the Blackfoot River. The finely bedded rocks are the Mount Sheilds Formation, and the rounded, grassy ridge is the diabase sill. These beds form the steep southwestern limb of the Wisherd syncline, the major fold of the Garnet Range. The Wisherd syncline plunges southeast and contains Paleozoic rocks east of here. Farther southeast, related structures involve middle Cretaceous shales of the Blackleaf Formation. The Blackfoot fault cuts through the vertical limb of the syncline and has about 5 kilometers of displacement here. Both the Wisherd syncline and the Blackfoot thrust are Laramide structures.

The sill forms an important structural marker in the Garnet Range and Jocko Mountains.

Continue east on Montana 200.

114.0 The Bonner Quartzite crops out on the prominent ridge on the left bank of the river.

114.5 MacNamara Formation crops out in roadcuts on the right.

115.2 **Stop 10: Southwestern limb of the Wisherd syncline (Garnet Range Formation).**

Park on the right side of the highway in the parking area just beyond the Blackfoot Tavern. Cross the highway and walk across the Blackfoot River on the swinging footbridge. Climb to the old railway berm and walk toward the left (downstream) about 1/4 mile.

The Garnet Range Formation contains broad channel sandstones and olive green micaceous shales, unlike the more characteristic planar-bedded Belt units. The deformational style is also different. This outcrop displays typical small-scale structures of the Garnet Range Formation. These structures are more intricate than those of other Belt formations. Note at least two nearly isoclinal synclines, with axial planar cleavage in pelitic layers. Sandstones are uncleaved, but bedding planes and fault planes in sandstones have prominent slickensides. These rocks were in the ductile deformation field for shale but in the brittle deformation field for sandstone when they were folded into the Wisherd syncline. Most of the mesoscopic structures define a unique fabric associated with the southeast-plunging fold system.

Return to vehicles and proceed east on Montana 200.

116.8 Talus blocks of Pilcher Formation are on the right.

118.0 Bridge over Blackfoot River.

118.6 Rest area is on the right. Cross the axial trace of the Wisherd syncline in Garnet Range Formation.

119.2 **Stop 11: Northeastern limb of the Wisherd syncline (Garnet Range Formation).**

Park on the right shoulder of the highway. Cross the highway and climb up to the abandoned railway berm.

This railroad cut displays the relationships among thrusting, folding, and cleavage. There are large northeast-facing kink folds of bedding and small northeast-directed thrusts. One thrust in a sandstone bed near the left (downstream) end of the outcrop passes into an asymmetric anticline in overlying shales. The anticline has axial plane cleavage consistent with cleavage throughout the Garnet Range, Jocko Mountains, and Boundary Divide.

Return to vehicles and proceed east on Montana 200.

121.0 Large roadcut in McNamara Formation. Turn around at large pulloff on the right and return to I-90.

Day 2: Bonner to Garrison.

Restart mileage at entrance to I-90 E, exit 109.

Mileage Description

0.0 Proceed east on I-90.

0.8 To the right is Milltown Dam. Excavation on

right side of dam exposes the sill in the Mount Shields Formation. Note several small extensional faults.

1.8 I-90 follows the trace of the Ninemile/Clark Fork fault zone for next 10 miles. Displacement across the fault diminishes toward the east.

6.1 The sill in the Mount Shields Formation crops out on the left. It is highly sheared, probably because of movement along the Clark Fork fault, which offsets the sill from the Milltown dam site on the south to here on the north.

8.7 The large quarry on the left is in the Bonner Formation. It is highly brecciated along the Clark Fork fault zone. Several steep southwest-dipping fault planes can be seen in the west quarry wall.

11.5 Town of Clinton. I-90 curves to the right and leaves the trace of the Clark Fork fault zone, which proceeds up Wallace Gulch. The Late Cretaceous Clinton stock forms the bed of Wallace Gulch. This granodiorite pluton crosscuts the Blackfoot fault.

15.0 I-90 crosses the Bearmouth thrust fault zone. Brecciated Bonner quartzite crops out along steep ridges on left.

15.8 Large roadcut on left reveals a rotated thrust fault.

20.8 Take Beavertail Road, exit 130.

21.0 Stop sign. The hills ahead and to the right are middle Eocene shallow intrusive rocks. These rocks postdate the thrust structures. Turn left and proceed under I-90.

21.4 Turn left and proceed up the hill on the secondary road.

21.7 Cambrian Hasmark Dolomite crops out on right, in footwall of Bearmouth thrust.

21.9 **Stop 12: Bearmouth thrust fault in Cramer Creek.**

Park at the topographic saddle.

Traverse east along Cramer Creek Road. The Bearmouth thrust is an important fault that overlies the Blackfoot thrust in the Garnet Range. It has exclusively brittle fabrics, in contrast with the Blackfoot thrust. At this locality, upper Missoula Group overrode Hasmark Dolomite. The cross-section (Figure 4) shows the Bearmouth thrust cutting across major folds in the footwall, which appear to be related to the Blackfoot thrust. The truncated folds are cleaved. There appears to have been out-of-sequence faulting in this area.

Continue up Cramer Creek Road. The road crosses down section in the footwall of Bearmouth thrust from Hasmark Dolomite to McNamara Formation.

23.4 West fork of Cramer Creek enters from left. The creek follows the trace of Clark Fork-Ninemile fault. Here the fault places Garnet Range Formation against McNamara Formation.

24.9 **Stop 13: Blackfoot thrust fault in Cramer Creek.**

Stop near the tailings heap and abandoned workings of the Linton Mine near the entrance to the narrow canyon of Cramer Creek, cut into Hasmark Dolomite.

The Blackfoot thrust here places McNamara Formation over a series of fault horses. The McNamara and Garnet Range Formations form the lower cliffs on the south side of the canyon. The highest horse is Pilcher Quartzite and Garnet Range Formation, exposed along the road. The next is Hasmark Dolomite, which forms the high cliffs on the north side of the canyon. The Hasmark horse overlies two horses of Hasmark through Jefferson Dolomite, which form ridges farther to the north.

The Blackfoot thrust climbed about 5 kilometers in the section in both the footwall and hanging wall between Stop 9 and here. In the hanging wall it climbed from the lower Missoula Group to the McNamara Formation; in the footwall, from the Mount Shields Formation to the Jefferson Dolomite. The fault was a mylonite zone at Stop 9; here it is a brecciated duplex structure. Consistent structural plunge to the southeast accounts for the climb in section and decrease in ductility between Stop 9 and here. Southeast from Cramer Creek, the Eocene Bearmouth volcanic rocks bury the Blackfoot fault. It may pass into a blind zone of disharmonic folding in the Maywood Formation (Figure 4).

Return to I-90.

28.9 Turn left on I-90 E.

29.7 Hill on the left is Eocene granite intruded by a picturesque basaltic dike.

31.1 Power station for the abandoned Milwaukie Railroad is on the right.

33.0 Eocene granite is on the left.

33.5 Hasmark Dolomite crops out on left side of I-90. The highway follows the approximate trace of the Bearmouth thrust.

34.5 At Milepost 136 is a good view east to the Bearmouth thrust, placing Bonner Formation over Hasmark Dolomite.

35.2 Nimrod Hot Spring tufa mound is on the left.

36.3 Bearmouth exit. Continue on I-90.

36.9 Bearmouth Chalet is on the left. McNamara Formation crops out on hill behind chalet.

39.2 The access road on the left side of the valley follows the approximate trace of the Bearmouth thrust fault, which here places MacNamara Formation over Mesozoic shales. The shales exposed in Little Bear Gulch are complexly folded in the southeast-plunging Mount Baldy syncline. The sparsely forested hill directly ahead across the Clark Fork River is a fault horse of Paleozoic rocks sandwiched between the Bearmouth thrust and the Mesozoic shales.

40.3 Bonner Formation in the hanging wall of the Bearmouth thrust crops out across the river to the left. Eocene Bearmouth volcanic rocks overlap the trace of the Bearmouth thrust and mask the Late Cretaceous structure for the next 1.5 miles.

40.8 Quarries across the river are in the Eocene Bearmouth volcanic rocks.

41.8 Bear Gulch gold placer dredge tailings across river.

42.3 Mississippian Madison Group limestones crop out in the deep canyon of the Clark Fork River which I-90 enters at this point.

42.8 **Stop 14: Bearmouth anticline.**

Stop on shoulder of I-90 in the deep canyon.

The large kink fold to the south in the Madison Group limestones is the Bearmouth anticline. This is one of a series of sharp-hinged folds of the Paleozoic carbonates which are disharmonic with broader folds in the underlying Belt Supergroup. The Blackfoot fault apparently passes into a blind decollement zone somewhere within the Paleozoic section (Figure 4).

Proceed east on I-90.

43.8 I-90 traverses upwards in the steeply tilted section into the Lower Cretaceous Kootenai Formation. Jurassic Ellis Group crops out across the Clark Fork River.

44.3 Long roadcuts across the river are in the middle limestone member of the Kootenai Formation.

45.3 The valley is in the Kootenai and Blackleaf Formations in the core of the box-shaped Mulky Gulch syncline.

46.3 **Stop 15: Mulky Gulch syncline.**

Stop along I-90 at east end of the outcrop. Walk south around the east end of the hill to a very large railroad cut.

This large railroad cut exposes the Mulky Gulch syncline in the upper part of the Kootenai Formation. The syncline plunges southeast, is overturned to the northeast, and has southwest-dipping cleavage. It is clearly part of the set of southeast-plunging structures we have traced from Paradise to here. The syncline is on the northeast flank of the Bearmouth anticline we viewed at Stop 14. The cliff to the southeast is Madison Limestone, exposed along a left lateral tear fault linked into the Bearmouth thrust. The tear fault cuts off the Mulky Gulch syncline. The east limb of the Mulky Gulch syncline is overturned to the southwest (Figure 4).

Continue east on I-90.

48.5 The Clark Fork River valley narrows here between the thrusted Madison Group to the south and the same rocks to the north, uplifted in a large southeast-plunging anticline.

51.2 Take Drummond exit 153. The letter "D" on the hillside ahead is in the core of an anticline outlined by the gastropod limestone member of the Kootenai Formation.

52.0 Turn left at the end of Drummond and pass under I-90. Proceed through pole plant on dirt road.

52.4 Take the high road to the left and cross the cattle guard.

52.6 **Stop 16: Gastropod limestone member of the Kootenai Formation.**

Stop at the quarry.

The large dipslope is held up by a bed of gastropod-rich limestone. There are also ostrocods, turtle plates, shark teeth, and bone fragments. A prominent spaced cleavage set intersects the bedding plane to form lines with a southeasterward plunge. The limestone is thrust over Blackleaf shales at the east end of the quarry. These shales have a pencil cleavage.

Return to I-90 at east end of Drummond.

53.2 Proceed east on I-90. To the right is the Flint Creek Range. These mountains contain high-level Late Cretaceous batholiths that intrude thrust and fold structures which curve to the south from Drummond. The Upper Cretaceous Flood and Dunkleburg Formations are on the left along the frontage road in southeast-plunging syncline. To the south, this syncline becomes nearly isoclinal and deforms a Late Cretaceous sill (Gwinn, 1961).

57.2 Tertiary and Quaternary deposits cover the Clark Fork "sag," a structural depression between the southeast-plunging folds of the Garnet Range and the north-plunging folds of the northern Flint Creek Range.

60.7 Jens exit. Stay on I-90. Cross through a thick section of Upper Cretaceous fine-grained clastic rocks poorly exposed on the left for the next 8 miles. This is the thickest Cretaceous section reported in Montana. The interval from the Kootenai Formation through the Carter Creek Formation is 3,577 meters thick (Gwinn, 1961), and the overlying Golden Spike Formation is 1,830 meters thick (Gwinn and Mutch, 1965).

63.2 Crossing the southeast-plunging Saddle Mountain anticline, one of several en echelon folds along the south limb of the Purcell anticlinorium.

66.2 Crossing the Carter Creek syncline. This is one of the major synclines along the south limb of the Purcell anticlinorium. It contains the Golden Spike Formation (Gwinn, 1961).

67.7 Late Cretaceous diorite sill is on the left.

68.2 Take Phosphate exit 168.

68.4 Turn right at intersection.

68.5 Cross bridge over Clark Fork River.

68.6 Cross railroad tracks and turn left.

68.8 **Stop 17: Golden Spike Formation "chaos beds."**

Park at the gate. Walk beside the tracks to the large outcrop. CAUTION! This is a busy track, and the trains approach quickly around the curve with little warning. Please stay well off the tracks.

Gwinn and Mutch (1965) called these the "chaos beds" of the Golden Spike Formation. They include pebbly mudstones and conglomerates with clasts derived from the Missoula Group, presumably from the western slab. Mackie (1986) suggested there were two periods of thrust uplift and detrital influx into the Golden Spike, based on sedimentary petrology. Gwinn and Mutch (1965) suggested the Golden Spike was derived from the thrust belt to the west and from the Elkhorn Mountains volcanic rocks to the east. Mackie (1986) mapped Elkhorn Mountains volcanic flows within the eastern part of the Golden Spike Formation. The Golden Spike Formation rests with angular unconformity upon the Carter Creek Formation (Gwinn, 1961).

The "chaos beds" contain broken formations, rolled and contorted beds, and matrix-supported boulders. Look for these at the west end of the outcrop. The matrix is cleaved.

Return to I-90.

69.4 Proceed east on I-90.

74.5 Take Garrison exit.

74.7 Turn right and take the frontage road back to the west, beside the railroad tracks.

75.2 **Stop 18: Large blocks in Golden Spike Formation "chaos beds."**

Park on left opposite large outcrop.

This outcrop of the Golden Spike "chaos beds" is thought to be at the same stratigraphic level as Stop 17 (Mackie, 1986). Very large blocks of bedded sandstone float in a pebbly mudstone and are probably derived from underlying Cretaceous units that may have been exposed by deformation related to emplacement of the western slab. The "chaos beds" are cut by a high-angle reverse fault at the west end of the outcrop.

SUMMARY

Stop 18 concludes the trip through the thrust system. This stop is about 25 kilometers higher in the Late Cretaceous structural section than Stop 1. Throughout this very thick section, this trip has traced a unifying cleavage surface, which changed from a biotite schistosity at depth to a phyllitic schistosity and then to a spaced, anastomosing pressure-solution cleavage at shallow levels. Cross-cutting relationships date the cleavage as Late Cretaceous. This trip climbs through the system along a composite cross-section that suggests the deformation was compartmentalized in three distinct levels, depending on depth, temperature, and rock composition.

ACKNOWLEDGEMENTS

Part of this route was used for the 1988 Tobacco Root Geological Society field conference. Bob Weidman reviewed and edited the guide for that trip. David Alt reviewed this manuscript. MONTS grants to Sears in 1986 and 1987 provided support for parts of the study.

Many thanks to V. Chamberlain, R. Breckenridge, and B. Bonnichsen for their work in organizing and editing this guidebook.

REFERENCES

Bally, A. W., 1984, Tectogenese et sismique reflexion: Bulletin Societe Geologic de France, v. 26, p. 279-286.

Cressman, E. R., 1985, The Prichard Formation of the lower part of the Belt Supergroup (Middle Proterozic), near Plains, Sanders County, Montana: U.S. Geological Survey Bulletin 1553, 64 p.

Desormier, W. L., 1975, A section of the northern boundary of the Sapphire tectonic block, Missoula and Granite Counties, Montana: University of Montana M.S. thesis, 64 p.

Emmons, W. H., and F. C. Calkins, 1913, Geology and ore deposits of the Philipsburg quadrangle, Montana: U.S. Geological Survey Professional Paper 78, 271 p.

Gwinn, V. E., 1961, Geology of the Drummond area, central-western Montana: Montana Bureau of Mines and Geology Geologic Map 4.

Gwinn, V. E., and T. A. Mutch, 1965, Intertongued Upper Cretaceous volcanic and nonvolcanic rocks, central-western Montana: Geological Society of America Bulletin, v. 76, p. 1125-1144.

Hall, F. W., 1968, Bedrock geology, north half of Missoula 30' quadrangle, Montana: University of Montana Ph. D. thesis, 253 p.

Harrison, J. E., 1972, Precambrian Belt basin of northwestern United States – Its geometry, sedimentation, and copper occurrences: Geological Society of America Bulletin, v. 83, p. 1215-1240.

Harrison, J. E., A. B. Griggs, and J. D. Wells, 1986, Geologic and structure maps of the Wallace 1x2 quadrangle, Montana and Idaho: Montana Bureau of Mines and Geology, Montana Atlas 4-A, scale 1:250,000.

Hyndman, D. W., 1980, Bitterroot dome-Sapphire tectonic block, an example of a plutonic-core gniess-dome complex with its detached suprastructure, *in* M. D. Crittenden, Jr., P. J. Coney, and G. H. Davis, editors, Cordilleran Metamorphic Core Complexes: Geological Society of America Memoir 153, p. 427-443.

———, 1986, Petrology of Igneous and Metamorphic Rocks: New York, McGraw-Hill, 786 p.

Hyndman, D. W., David Alt, and J. W. Sears, 1988, Post-Archean metamorphic and tectonic evolution of western Montana and northern Idaho, *in* W. G. Ernst, editor, Metamorphism and Crustal Evolution of the Western United States: Rubey Volume VII, New Jersey, Prentice-Hall, p. 332-361.

Kauffman, M. E., 1961, Geology of the Garnet-Bearmouth area, western Montana: Montana Bureau of Mines and Geology Memoir 39, 40 p.

Lonn, Jeffrey, 1986, Geology of the Tarkio area, Montana: University of Montana M.S. thesis, 51 p.

Mackie, T. L., 1986, Tectonic influences on the petrology, stratigraphy and structures of the upper Cretaceous Golden Spike Formation, central-western Montana: Washington State University M.S. thesis, 132 p.

Maxwell, J. C., 1965, Structure map and cross sections, southeast Drummond area, *in* R. W. Fields and Warren Shepard, editors, Geology of the Flint Creek Range, Montana: Billings Geological Society, 16th Annual Field Conference, Map.

Minnich, G. W., 1984, The timing of metamorphism relative to rock fabric in the Silver Hill Formation near the Garnet stock: University of Montana B.S. thesis, 11 p.

Mudge, M. R., R. L. Earhart, J. W. Whipple, and J. E. Harrison, 1982, Geologic and structure map of the Choteau $1^{o}x2^{o}$ quadrangle, western Montana: U.S. Geological Survey, Miscellaneous Investigations, Map I-1300, scale 1:250,000.

Myers, S. A., 1986, Significance of the chlorite breccia zone, Bitterroot dome, western Montana: University of Montana M.S. thesis, 72 p.

Nelson, W. H., and J. P. Dobell, 1961, Geology of the Bonner quadrangle, Montana: U.S. Geological Survey

Bulletin 111-F, p. 189-235.

Nold, J. L., 1968, Geology of the northeastern border zone of the Idaho batholith, Montana and Idaho: University of Montana Ph. D. thesis, 159 p.

Norwick, S. A., 1972, The regional Precambrian metamorphic facies of the Prichard Formation of western Montana and northern Idaho: University of Montana Ph. D. thesis, 129 p.

Obradovich, J. D., and Z. E. Peterman, 1968, Geochronology of the Belt Series, Montana: Canadian Journal of Earth Science, v. 5, p. 737-747.

Price, R. A., 1981, The Cordilleran thrust and fold belt in the southern Canadian Rocky Mountains, *in* K. R. McClay and N. J. Price, editors, Thrust and Nappe Tectonics: Geological Society of London Special Publication 9, p. 427-448.

Ruppel, E. T., C. A. Wallace, R. G. Schmidt, and D. A. Lopez, 1981, Preliminary interpretation of the thrust belt in southwest and west-central Montana and east-central Idaho: Montana Geological Society Field Conference Guidebook, p. 139-159.

Sears, J. W., 1985, Middle Eocene tuff from the Missoula basin: Northwest Geology, v. 14, p. 33-36.

———, 1988a, Changes in Late Cretaceous structural styles between Ninemile and Drummond, Montana, *in* R. M. Weidman, editor, Guidebook of the Greater Missoula Area: Tobacco Root Geological Society, 13th Annual Field Conference, p. 42-54.

———, 1988b, Two major thrust slabs in the west-central Montana Cordillera, *in* C. J. Schmidt and W. J. Perry, Jr., editors, Interaction of the Rocky Mountain Foreland and the Cordilleran Thrust Belt: Geological Society of America Memoir 171, p. 165-170..

Sears, J. W., L. M. Johnson, B. C. Geiger, and W. C. Brandon, 1988, Southwest Montana thrust belt: Bannack to Melrose, *in* P. K. Link and W. R. Hackett, editors, Guidebook to the Geology of Central and Southern Idaho: Idaho Geological Survey Bulletin 27, p. 305-319.

Thomas, M. B., 1987, Structural geology along part of the Blackfoot fault system near Potomac, Missoula County, Montana: University of Montana M.S. thesis, 52 p.

Turcotte, D. L., and G. Schubert, 1982, Geodynamics: New York, John Wiley & Sons, 450 p.

Wallace, C. A., R. G. Schmidt, M. R. Waters, David Lidke, and A. B. French, 1981, Preliminary geologic map of the Butte 1 by 2 degree Quadrangle, central Montana: U.S. Geological Survey Open-File Report 81-1030, scale 1:250,000.

Watson, I. A., 1984, Structure of the Rattlesnake area, west-central Montana: University of Montana M.S. thesis, 55 p.

Weiss, C. P., 1987, A comparison of the thermal profiles of two adjacent thrust plates in western Montana: University of Montana M.S. thesis, 76 p.

Wells, J. D., 1974, Geologic map of the Alberton quadrangle, Missoula, Sanders and Mineral Counties, Montana: U.S. Geological Survey Geologic Quadrangle Map GQ-1157.

Winston, Don, and Jeffrey Lonn, 1988, Structural geology and Belt stratigraphy of the Tarkio area, Montana, *in* R. M. Weidman, editor, Guidebook of the Greater Missoula Area: Tobacco Root Geological Society 13th Annual Field Conference, p. 1-34.

The Cambrian System of Northern Idaho and Northwestern Montana

John H. Bush [1]

INTRODUCTION

Cambrian rocks are exposed at several locations in northern Idaho and northwestern Montana (Figure 1). In northern Idaho, the most extensive outcrops of Cambrian rocks are along the southeast side of Pend Oreille Lake near the town of Lakeview. These Cambrian strata occur in fault-bounded blocks surrounded by rocks of the Precambrian Belt Supergroup and by Cretaceous granodiorite. The sequence at Lakeview (Figure 2) is complexly folded, faulted, and locally metamorphosed. From the base upward it consists of the Gold Creek Quartzite (122 m), the Rennie Shale (30 m), and the Lakeview Limestone (587 m).

In northwestern Montana, between the towns of Thompson Falls and Libby, Cambrian rocks are preserved in a northwest-striking syncline referred to as the Libby Trough (Figure 3). The basal unit is the Flathead Sandstone (8 m), which is overlain by the Gordon Shale (75 m). The Gordon Shale is overlain by a carbonate unit described by Aadland (1985) as the Fishtrap Formation (895 m).

In addition to the Lakeview and Libby Trough areas, three small exposures (Figure 1) of Cambrian rocks have been identified in northern Idaho and northwestern Montana (Hayden and Bush, 1988). One outcrop of dolostone (34 m) occurs 7.8 kilometers northeast of Moyie Springs, Idaho. Another outcrop of dolostone (170 m) occurs on Lime Butte, 11 kilometers south of Troy, Montana. A third outcrop near Heron, Montana, consists of argillaceous limestone (102 m).

Platform evolution and paleogeography can be outlined by regional correlations to the surrounding areas. Early Middle Cambrian ramp deposition began with nontectonic highs in northwestern Montana and distally steepened areas in northeastern Washington (Bush and Hayden, 1987). After transgression, upward-shallowing

[1] *Department of Geology and Geological Engineering, University of Idaho, Moscow, Idaho 83843.*

Bush, J. H., 1989, The Cambrian system of northern Idaho and northwestern Montana, *in* V. E. Chamberlain, R. M. Breckenridge, and Bill Bonnichsen, editors, Guidebook to the Geology of Northern and Western Idaho and Surrounding Area: Idaho Geological Survey Bulletin 28, p. 103-121.

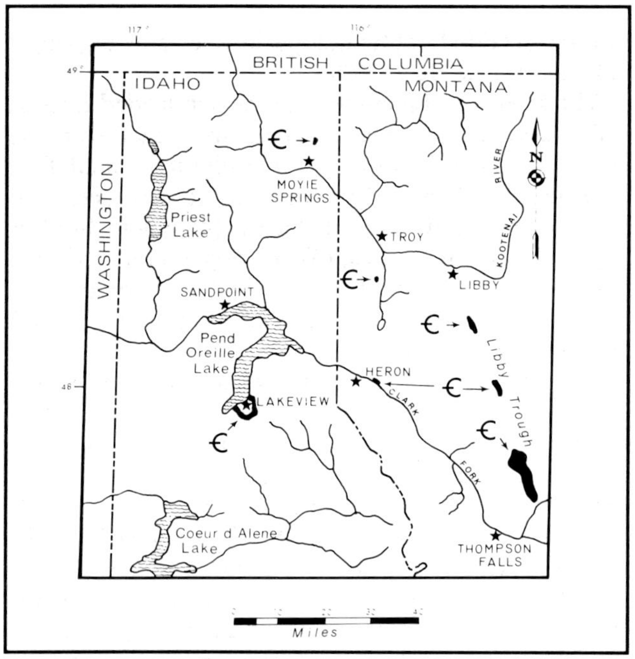

Figure 1. Location map showing Cambrian outcrops in northern Idaho and northwestern Montana.

Middle Cambrian carbonates formed a peritidal, algal-shoal complex that extended from central Montana to northeastern Washington (Bush and Fischer, 1981). Growth of the complex was at times reduced by clastic influx from both the Lemhi Arch in central Idaho and the inner detrital belt, but the complex remained an important paleogeographic feature until at least the end of the Cambrian. For most of the Middle and Late Cambrian, the complex separated an intrashelf basin from the outer ramp. During the Late Cambrian, below-wave-base carbonate and clastic deposition returned to onlap the peritidal carbonates in northeastern Washington. The expansion and contraction of the algal-shoal complex can be related to the Grand Cycles of Aitken (1978), which are useful for regional correlations and the construction of depositional models. The correlation of Cambrian sequences to Grand Cycles and the development of the algal-shoal complexes are emphasized throughout this guide.

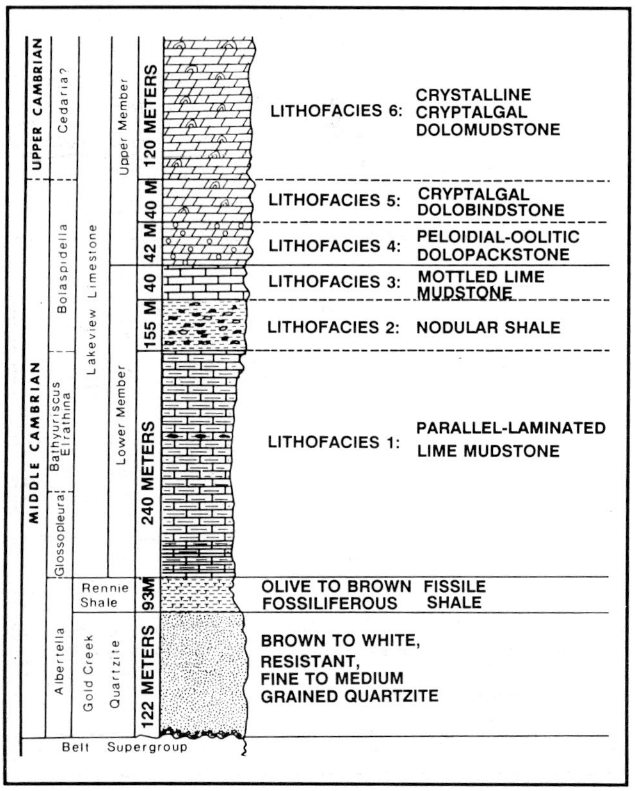

Figure 2. Generalized stratigraphic section of the Cambrian strata near Lakeview, Idaho.

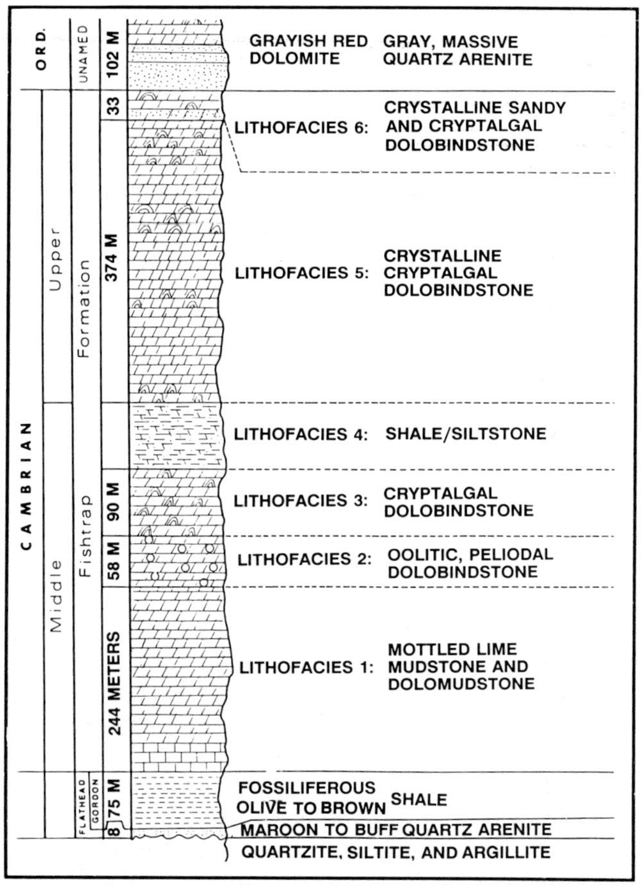

Figure 3. Generalied stratigraphic section of the Cambrian and Ordovician strata in the Libby trough of northwestern Montana.

STRATIGRAPHIC UNITS ALONG PEND OREILLE LAKE

Gold Creek Quartzite and Rennie Shale

The Gold Creek Quartzite was named for its exposures on north and south Gold Creeks near the town of Lakeview, Idaho (Sampson, 1928). The quartzite rests unconformably on the lower part of the middle member of the Wallace Formation of the Belt Supergroup (Harrison and Jobin, 1965). The rock consists primarily of massive cross-bedded quartz arenite with minor interbedded shale, and a basal quartz cobble and pebble conglomerate.

The Rennie Shale was named by Sampson (1928) for exposures on the south side of Packsaddle Mountain northeast of Lakeview. The unit consists of fossiliferous, olive to brown, fissile shale with interbedded nodules of fossiliferous lime mudstone. The unit is soft and easily eroded; outcrops are rare. Collections of trilobites, one from near the base and one from the upper part, were examined and both collections correspond with the *Albertella* zone (Harrison and Jobin, 1965). The Rennie Shale is considered to be the lithologic equivalent of the Gordon Shale and Wolsey Shale of western Montana.

The only known complete exposure of the Rennie Shale occurs along a tributary of North Gold Creek in the Packsaddle Mountain quadrangle, about 3.3 kilometers southeast of Packsaddle Mountain. The exposures are moss covered and can be sampled only by standing in the stream. Harrison and Jobin (1965) mapped the Rennie Shale at this locality along the northern one-half boundary of sections 29 and 30, T. 54 N., R. 1 E.

Lakeview Limestone

Stratigraphic and structural controls of the Lakeview Limestone are complex. The most continuous, unmetamorphosed outcrops occur along the northern part of Gold Creek just south of the town of Lakeview. Composite sections along Gold Creek indicate that the Lakeview Limestone is about 600 meters thick. The Lakeview Limestone is in conformable contact with the underlying Rennie Shale. The upper contact is eroded and overlain by Pleistocene glacial debris.

The lower Lakeview Limestone (335 m) consists primarily of dark gray, laminated, argillaceous, pyritic, fossiliferous lime mudstone with interbedded calcareous shale and nonlaminated lime mudstone. The upper Lakeview Limestone (252 m) consists primarily of light gray to tan, nonpyritic, nonfossiliferous dolomudstone with interbedded oolitic dolopackstones and cryptalgal dolobindstones (Bush, 1977).

The Lakeview Limestone can be subdivided into six major lithofacies (Figure 2). These are named from the base upward as follows: (1) parallel-laminated lime mudstone lithofacies; (2) nodular shale lithofacies; (3) mottled lime mudstone lithofacies; (4) peloidal-oolitic dolopackstone lithofacies; (5) cryptalgal dolobindstone lithofacies; and (6) crystalline cryptalgal dolomudstone lithofacies. The lower Lakeview consists of lithofacies 1-3 and the upper Lakeview of lithofacies 4-6.

Lithofacies 1 is 240 meters thick and consists primarily of dark gray, pyritic, sparsely fossiliferous, parallel-laminated lime mudstone interbedded with massive, black, lime mudstone. One interval, 150 meters above the base, contains lenticular to ellipsoid concretions of black, petroliferous lime mudstone with a trilobite-rich "hash" on a few surfaces. The nodular shale lithofacies 2 of the lower Lakeview Limestone is 55 meters thick and consists of a dark gray to tan, fossiliferous, calcareous, laminated shale interbedded with medium gray, fossilferous nodules of lime mudstone. Overlying lithofacies 2 is the mottled lime mudstone lithofacies (3) about 40 meters thick. The mottling occurs in black, massive lime mudstone that weathers to a light gray color.

The lower Lakeview Limestone, which includes lithofacies 1, contains faunas of the *Glossopleura* and *Bathyuriscus-Elrathina* zones (Lochman-Balk, 1971). Robison (1964) states that some of the faunas of the lower Lakeview Limestone belong to the *Bolaspidella* zone. Motzer (1980), using Robison's (1976) revision of biostratigraphic zones (Figure 4) for the Great Basin, placed the boundary between lithofacies 1 and 2 to approximate the boundary between the underlying *Oryc-*

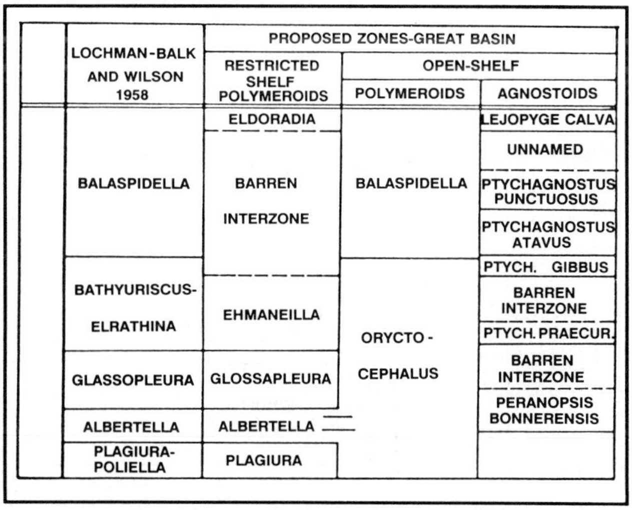

Figure 4. Comparison of Middle and Late Cambrian biostratigraphic zones of Lochman-Balk (1972) with those proposed by Robison (1976).

tocephalus and overlying *Bolaspidella* zones for open shelf polymeroid faunas.

The three lithofacies of the upper Lakeview Limestone are all nonfossiliferous and dolomitized. Peloidal, oolitic dolopackstones of lithofacies 4 are 42 meters thick and contain intercalated pisolitic grainstones and crystalline dolomudstones. Lithofacies 5 is 40 meters thick and consists of flat-lying cryptalgal dolobindstones with interbedded dolomudstones. Tepee structures, stromatolites, mudcracks, and raindrop impressions are present in lithofacies 5. Colors range from gray to purple; green units are most abundant. Lithofacies 6 is 170 meters thick and consists of a white, coarse, crystalline dolostone interbedded with a very dense, gray crystalline dolomudstone. However, flat-lying cryptalgal laminations in the gray crystalline dolomudstones are present on a few weathered surfaces.

Lithofacies 1 and 2 were deposited in low current, "deep shelf," anaerobic environments. Lithofacies 3 represents a stable, aerobic, subtidal environment. Fauna in the sediment proliferated and burrowed to produce the mottling. Lithofacies 4 was probably deposited as mud banks near tidal shoals in the intertidal zone where ooids could be carried in and deposited during storms.

Continued shallowing of the carbonates produced the supratidal rocks of lithofacies 5. The sedimentary structures in lithofacies 5 indicate that evaporitic conditions existed which were in part responsible for the extensive, early diagenetic dolomitization of the upper Lakeview. Lithofacies 6 was probably deposited in intertidal-supratidal conditions, although intense dolomitization and recrystallization have made interpretation difficult.

In summary, the Lakeview Limestone represents continuing upward-shallowing environments from "deep shelf" for lithofacies 1 to intertidal-supratidal environments for lithofacies 5 and 6.

STRATIGRAPHIC UNITS IN THE LIBBY TROUGH

Flathead Sandstone and Gordon Shale

In the Libby Trough area, the Flathead Sandstone can easily be subdivided into two units. The basal unit (3-5 m) consists of a maroon and buff, banded, well-indurated, fine- to medium-grained quartz arenite. Parallel laminated units dominate, but cross-beds are also common. Locally present are flat-crested, asymmetrical ripples, parting lineations, and bioturbation features. Apgar (1986) identified the burrows in the basal unit that increase upward in abundance as *Monocraterion* and *Skolithos*.

The upper unit (2-3 m) consists of a light-colored, coarse-grained to conglomeratic quartz arenite. The unit is cross-bedded with conglomeratic lenses, in places forming the foresets of the cross-beds. Locally, the unit is extensively burrowed. Apgar (1986) noted that the basal portion of this unit contains horizontal traces of *Thalassinoides*, *Cruziana*, and *Pharolites* or *Palaeophycus* and near the top abundant vertical traces of *Monocraterion*, *Skolithos*, *Diplocraterion*, *Arenicolites*, and *Corophioides*.

The Gordon Shale in the Libby Trough ranges from 26 to 75 meters in thickness and is characterized by its "cornflake-like" consistency. The shales are generally laminated and slightly fossiliferous with trilobite fragments the most common. At several outcrops, a glauconitic sandstone (0.5-2 m) is interbedded with the shales of the lower Gordon. The sandstones are cross-bedded and intensely bioturbated.

The upper Gordon Shale is gradational with the overlying Fishtrap Formation. Aadland (1979) placed the contact where the lime mudstones predominate over the shales. O'Malley (1985) placed the Gordon-Fishtrap contact at the base of the first major exposure of black lime mudstone.

The basal Gordon Shale is in the *Palgiura-Kochaspis* faunal zone, the middle Gordon in the *Albertella* zone, and the upper Gordon in the *Glossopleura* zone (O'Malley, 1985). Earlier work by Keim and Rector (1964) placed the Flathead-Gordon contact near the *Albertella* zone. The Gordon Shale has been correlated with the Wolsey Shale in southwestern Montana. Keim and Rector (1964) and Aadland (1979) used the term "Wolsey" for the fossiliferous shales in the Libby Trough. Aadland (1985), O'Malley (1985), and this field guide follow Gale (1934) who first reported the shale unit in the Libby Trough and used the term "Gordon Shale."

Fishtrap Formation

Keim and Rector (1964) were the first to describe a thick gray dolomitic unit that conformably overlies the Gordon Shale in the Libby Trough. They referred to the unit informally as the "Fishtrap dolomite." Aadland (1979) measured and described the sequence, which is about 900 meters thick. He retained the informal name but later used the name "Fishtrap Formation" (Aadland, 1985). Paleontological data from the Gordon Shale and overlying Ordovician rocks indicate that the Fishtrap ranges from the Middle Cambrian (*Glossopleura* zone) to the latest Cambrian.

Aadland (1979, 1985) measured five sections in the Libby Trough and subdivided the Fishtrap Formation into several lithofacies. For the purposes of this discussion, these lithofacies have been modified and labeled from the base upward as 1 through 6 and named as follows: (1) mottled lime and dolomudstone lithofacies; (2)

oolitic, peloidal dolograinstone lithofacies; (3) cryptalgal dolobindstone/mudstone lithofacies; (4) shale/siltstone lithofacies; (5) crystalline cryptalgal dolobindstone lithofacies; and (6) crystalline sandy dolostone and cryptalgal dolobindstone lithofacies (Figure 3). With the exception of the basal 35 meters, the entire unit is dolomitized.

Lithofacies 1 is about 250 meters thick and can be subdivided into four units. The basal 35 meters consists of thin to very thin beds of black, sparsely fossiliferous, argillaceous, mottled lime mudstone. These lime mudstones are overlain by 50 meters of mottled, pelitic, and intraclastic dolowackestones-packstones that are, in turn, overlain by a 25-meter-thick dolomitic shale and 121 meters of mottled, locally intraclastic, pelitic dolomudstones/ wackestones (Aadland, 1979). Color mottling dominates the sequence that owes its origin to organic churning of the sediments.

Lithofacies 1 was deposited in offshore subtidal areas, generally below the fair weather wave base, but not below the storm wave base. Aadland (1985) considered the lowermost lime mudstone unit to have been deposited close to the shoreline where it was influenced by silt and clay from the craton. The shale interval may represent a change in the rate of transgression or local reactivation of highs to bring in the extra terrigenous sediments. The overlying mottled dolomudstones were deposited seaward of the lime mudstone (Aadland, 1985).

Lithofacies 2 averages 58 meters in thickness and is characterized by three oolitic, peloidal dolograinstone beds that are separated by thinner, mottled dolomudstone beds. The overlying lithofacies 3 was named for the abundant flat-lying cryptalgal dolobindstones, although intraclastic and peloidal dolomudstones and dolowackestones are also present.

A comparison with modern analogs indicates that lithofacies 2 was deposited in shallow shoal environments that had vigorous wave or current activity on intertidal areas on or at the edge of the shelf. Locally the shoals were interlaced with shallow subtidal basins in which the bedded and bioturbated dolomudstones accumulated. Lithofacies 3 represents further shallowing to supratidal environments with subaerial exposure and penecontemporaneous dolomitization.

Lithofacies 4 consists primarily of dolomitic silty shale and clayey siltstone. The lithofacies is thickest on the southern end of the Libby Trough, where, according to Aadland (1979) it attains a thickness of 90 meters. The lithofacies is composed predominantly of thin beds of silty shale and clayey siltstone with papery partings. Dispersed among the silty shale, and clayey siltstone beds are very thin- to thin-bedded argillaceous dolostone interbeds. This unit contains interlaminated terrigenous muds and silts and a variety of sedimentary structures such as flaser bedding, wave laminations, and parallel to lenticular bedding.

Aadland (1985) attributes the features of lithofacies 4 to deposition similar to those described for "mixed flat" and "mud flat" on modern tidal flats by Reineck (1972). He also notes that the silt and dolostone beds scattered throughout lithofacies 4 display similarities to the dololaminate unit in the Mauv Limestone described by Wanless (1975), who proposed that the unit records deposition in environments similar to the channeled tidal belt on the west flank of Andros Island in the Bahamas.

Lithofacies 5 consists of 374 meters of flat-lying cryptalgal dolobindstones interbedded with dolomudstones, dolowackestones, and crystalline dolostones. Features indicate a return to carbonate supratidal conditions that were interlaced with intertidal and subtidal basins.

The uppermost Fishtrap (lithofacies 6) consists of crystalline sandy dolostones and crystalline dolomudstones that attain a thickness of 33 meters. Aadland (1979) still recognized a few primary features such as laminations, pellets, intraclasts, and mottling. Flat-lying cryptalgal laminations are present in a few places on weathered surfaces. The sandy portions occur at the base of lithofacies 6, and cryptalgal features are more common near the top. The sand in the lower portions reduced algal growth as subtidal conditions returned. Supratidal and intertidal conditions prevailed during the deposition of the upper part of lithofacies 6 as environments shallowed and algal growth was again common.

Unnamed Ordovician Strata

Conformably overlying the Fishtrap Formation near Wee Peak in the southernmost block of the Libby Trough is an unnamed sequence of quartz arenite (65 m) and upper dolostone (37 m). The quartz arenites are massive, very mature, medium grained, and silica cemented. The dolostones are grayish red and locally intraclastic; in places they contain poorly preserved algal heads. Most of the dolostones are medium to coarsely crystalline, but very fine dolomudstones are also present.

Early Ordovician conodonts (Bush and others, 1985) were recovered from two intervals in the dolostones. This is the only reported Ordovician outcrop in northwestern Montana and northern Idaho.

REGIONAL CORRELATIONS, GRAND CYCLES, AND DEPOSITIONAL SETTING

Introduction

The Cambrian sequences at Lakeview, Idaho, and in the Libby Trough area of Montana have been compared with the Cambrian System in northeastern Washington and central and southwestern Montana (Lochman-Balk, 1972; Aadland, 1979, 1985; Motzer, 1980; Bush and others, 1980; Bush and Fischer, 1981; Bush and Hayden,

	NORTH EASTERN WASHINGTON	LAKEVIEW IDAHO	LIBBY TROUGH MONTANA	LEWIS CLARK RANGE MONTANA	SOUTH WESTERN MONTANA
LOWER ORDOVICIAN	LEDBETTER		UNNAMED SEQUENCE		
UPPER CAMBRIAN	METALINE FORMATION: UPPER		FISHTRAP FORMATION: LITHOFACIES 6		RED LION
			LITHOFACIES 5	DEVILS GLEN	PILGRIM
MIDDLE CAMBRIAN	MIDDLE	LAKEVIEW FORMATION: LITHOFACIES 6	LITHOFACIES 4	SWITCH BACK	PARK
		LITHOFACIES 5	LITHOFACIES 3	STEAMBOAT	MEAGHER
		LITHOFACIES 4	LITHOFACIES 2	POGODA	
		LITHOFACIES 3	LITHOFACIES 1	DEARBORN	
				DAMNATION	
	LOWER	LITHOFACIES1-2	GORDON	GORDON	WOLSEY
		RENNIE GOLD CREEK	FLATHEAD	FLATHEAD	FLATHEAD
LOWER CAMBRIAN	MAITLEN GYPSY	———	———	———	———

Figure 5. Generalized Middle and Late Cambrian correlation and nomenclature chart for northeastern Washington, northern Idaho, and western Montana.

1987). Figure 5 is a generalized correlation and nomenclature chart of the Cambrian from northeastern Washington to southwestern Montana. Bush and Fischer (1981) and Bush and Hayden (1987) used Grand Cycles in their comparisons as defined by Aitken (1978) from Cambrian and Lower Ordovician strata of the southern Canadian Rocky Mountains.

Aitken (1966, 1978) defined "Grand Cycles" as depositional cycles, each of which is represented by 90 to 600 meters of strata, one or more formations, and two or more biostratigraphic zones. Each cycle has an abrupt basal contact and consists of a lower, shaley half-cycle that is gradationally overlain by a carbonate half-cycle. Aitken (1978) notes that no two Grand Cycles are identical because the individual components that make up each cycle vary with differences in paleogeographic location. Therefore, correlations across depositional belts are difficult. However, Aitken (1981), Palmer (1981), and Chow and James (1987) demonstrate that regional correlations of Cambrian Grand cycles may be possible. The concept of Grand Cycles has also been applied to Cambrian strata in the Great Basin (Palmer and Halley, 1979; Mount and Rowland, 1981) and in the Northwest Territories (Fritz, 1975).

Three principal Cambrian shelf environments of the Cordilleran miogeocline, referred to as the inner detrital, middle carbonate, and outer detrital belt, have been established by Palmer (1960, 1971), Robison (1960), and Oriel and Armstrong (1971). These three principal environments are applicable to northwestern Montana and northern Idaho and have been used in those areas by numerous workers including Lochman-Balk (1972) and Aitken (1978). The inner detrital belt contains sandy and shaley sediments deposited along the western shore of North America. Clastics derived from the eastern craton accumulated in shoreface and bay environments. The middle carbonate belt consists of limestone or dolostone deposited in very shallow water offshore from the inner detrital belt. Typical lithologies of the middle carbonate belt include shallow water limestone and dolostone such as algal bindstone, mottled mudstone, bioclastic, oolitic, oncolitic wackestone and grainstone, and flat-pebble conglomerate. Offshore from the carbonate belt, the outer detrital belt consists of deposits of the outer shelf, continental slope, and basin sediments. Siltstone, shale, and dark limestone are characteristic. Deposits of the outer detrital belt are derived of sediment transported from the inner detrital belt and from material that slumps along the seaward edges of the middle carbonate belt.

A comparison of Aitken's (1966, 1978) Grand Cycles with the principal shelf environments shows that the shaley half cycle represents the inner detrital belt while the carbonate half-cycle represents the middle carbonate belt. The abrupt basal contact of the Grand Cycle is thought to be nearly synchronous (Aitken, 1966, 1978). The initiation of a Grand Cycle results from a relatively rapid marine transgression, which causes a sudden increase in the water depth and the supply of terrigenous material. The contact of the shaley half-cycle and carbonate half-cycle is diachronous, caused by a stabilization of the transgression which in turn caused a gradual shoaling in clear water and the expansion of the middle carbonate belt over the inner detrital belt (Aitken, 1966, 1978). Robison (1960) and Lochman-Balk (1971, 1972) contrast Aitken's ideas and consider the carbonates to have formed during transgression and the clastics to have been deposited during regional regression and emergence. Chow and James (1987) suggest that eustatic sea level changes appear to be the most feasible explanation of Cambrian Grand Cycles. Indeed, the Cambrian was a period of widespread sea-level rise (Holland, 1971). Several workers have proposed that differential rates were the cause of the cycles (Aitken 1978, 1981; Palmer and Halley, 1979; and Mount and Rowland, 1981). The basal half of a Grand Cycle represents a rapid rise in relative sea level, whereas the upper half-cycle represents a decrease in the rate of sea level rise (Chow and James, 1987).

Grand Cycles as defined by Aitken (1966, 1978) are easily recognized in southwestern Montana (Figure 6). A basal cycle is represented by the Wolsey to Meagher sequence (early Middle to mid-Middle Cambrian), a second cycle by the Park to Pilgrim sequence (late Middle to early Late Cambrian), and a third cycle by the Red Lion Formation (middle Late to late Late Cambrian). In northwestern Montana and northern Idaho, the tops of cycles are difficult to recognize because deposition was dominated by middle carbonate belt environments, away from the influence of the inner detrital belt. However, detailed lithologic data and paleontological data can be used to determine approximate cycle boundaries at several localities.

In the Libby Trough, the Gordon Shale and the basal Fishtrap Formation to the top of lithofacies 3 represents the basal Grand Cycle. The end of the same cycle is represented in the Lakeview Limestone by lithofacies 5 that contains features indicating subaerial exposure. Lithofacies 4 of the Fishtrap Formation represents the clastic portion of the second Cambrian Grand Cycle, and lithofacies 5 contains at least part, if not all, of the carbonate half of the cycle. The clastic portion of the second cycle is missing in the Lakeview Limestone, and the entire cycle may be represented by lithofacies 6. The beginning of the third cycle is difficult to determine in the Libby Trough because of the lack of a dominant algal unit in the top of lithofacies 5. Tentatively, the basal sandy dolostone of lithofacies 6 is interpreted to represent the clastic portion of the third cycle. The end of the third cycle is easily recognized as the contact between the Ordovician quartz arenites and lithofacies 6 of the Fishtrap Formation. The thin nature of the third Grand Cycle (33 m) in the Libby Trough makes the selection of the contact between the second and third Grand Cycle suspect. The Ordovician quartz arenite to carbonate sequence clearly represents a fourth cycle.

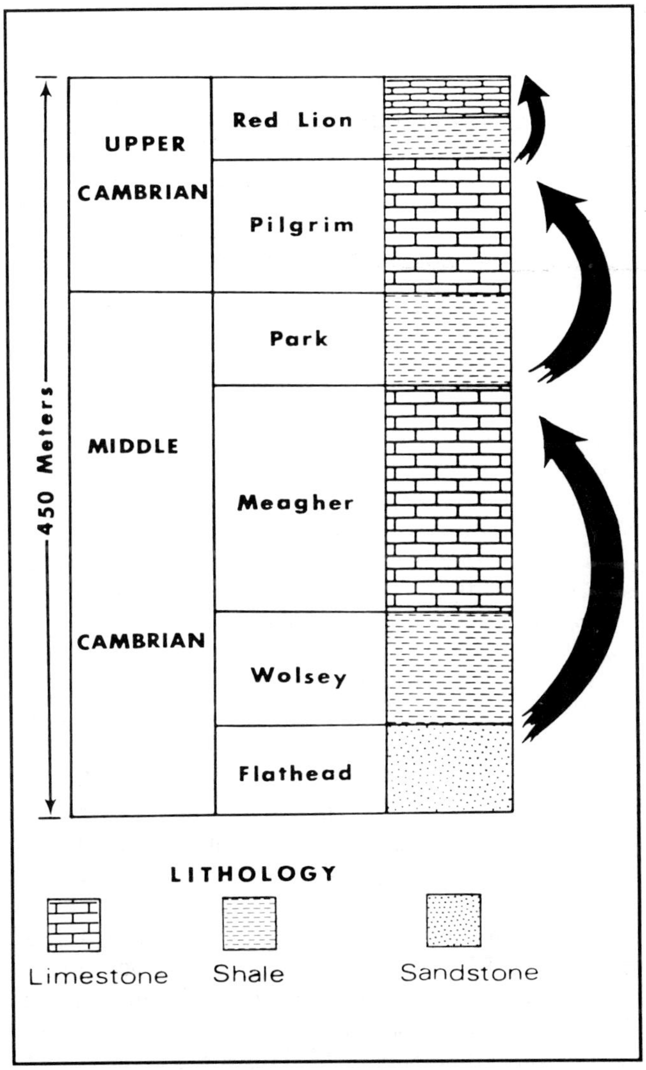

Figure 6. Generalized stratigraphic section for the Cambrian of southwestern Montana with Grand Cycle interpretations illustrated with upward-curving arrows. Modified from Healy (1985).

Depositional Events

Cambrian deposition in northern Idaho and northwestern Montana began with an early Middle Cambrian transgression towards the craton with the deposition of the Gold Creek, Rennie, and lowermost Lakeview. Figure 7 illustrates this setting with the deposition of distal argillaceous carbonates in northeastern Washington (lower Metaline) and bioturbated carbonates in northern Idaho (Lakeview lithofacies 3), while the sand and mud of the inner detrital belt were deposited over a preexisting Cambrian highland in northwestern Montana.

The first major Cambrian transgression into Montana was relatively rapid. Upon stabilization and the clearing of the water, shallowing and carbonate production occurred on a regional basis. Lithofacies 3 to 5 of the Lakeview and lithofacies 1 to 3 of the Fishtrap represent the carbonate half of this first Middle Cambrian Grand Cycle. The end of the cycle correlates to the top of the Meagher Formation in southwestern Montana, and to some horizon in the lower part of the middle Metaline Formation in northeastern Washington, and to the top of the Steamboat Limestone in the Lewis-Clark Range of west-central Montana.

The regional production of carbonates at the end of the first cycle formed an extensive, peritidal, algal-shoal complex that prograded basinward and landward causing shallow carbonates to overlie more distal carbonates at Lakeview, Idaho, and Metaline Falls, Washington (Figure 8). The same event produced an intrashelf basin between the algal shoals and the cratonic shoreline. At its maximum width, the complex extended from Helena, Montana, to Metaline Falls, Washington. The algal shoals were part of a much larger complex that extended in length from Edmonton, Canada, to southern Utah (Aitken, 1978). The development of the Cambrian algal shoals in northern Idaho and northwestern Montana has been documented by Lochman-Balk (1971), Aadland (1979, 1985), Bush and others (1980), Martin and others (1980), and Bush and Fischer (1981).

Migration of the algal-shoal complex and changes in its size influenced the following:

1. Evaporation, algal growth, and early penecontemporaneous dolomitization on and beneath the shoal complex.
2. Salinity and distribution of normal marine fauna in the intrashelf basin.

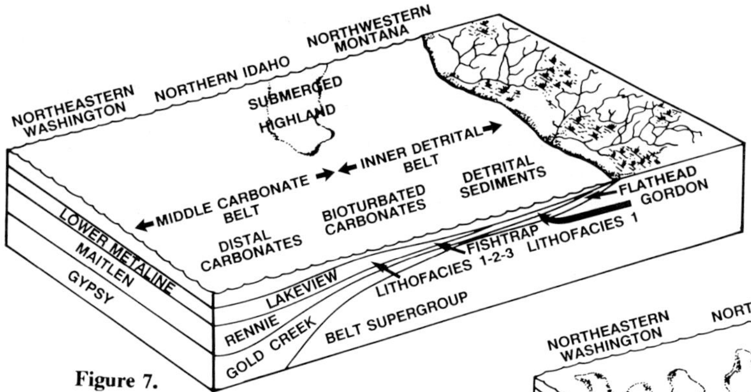

Figure 7.

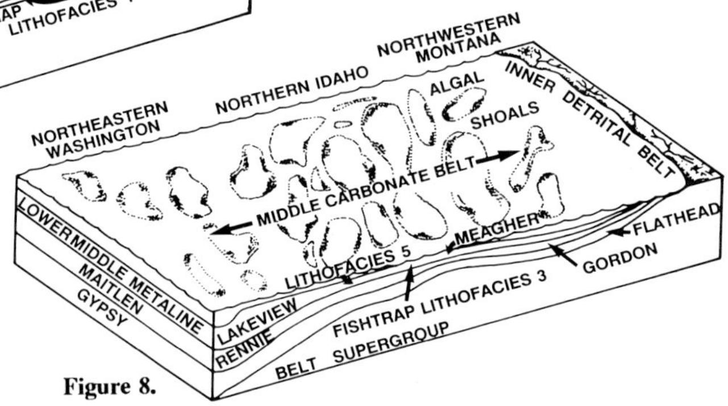

Figure 8.

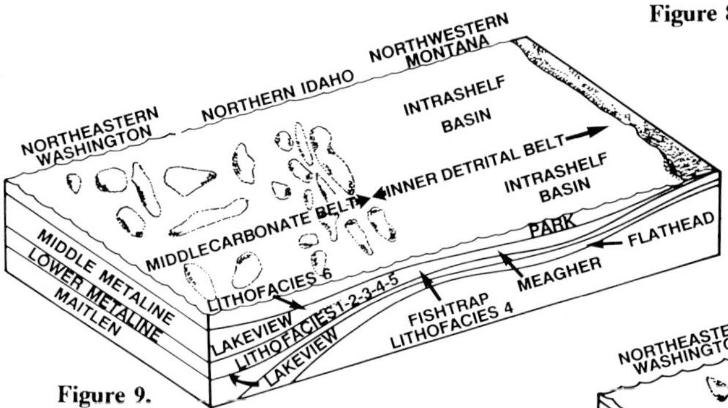

Figure 9.

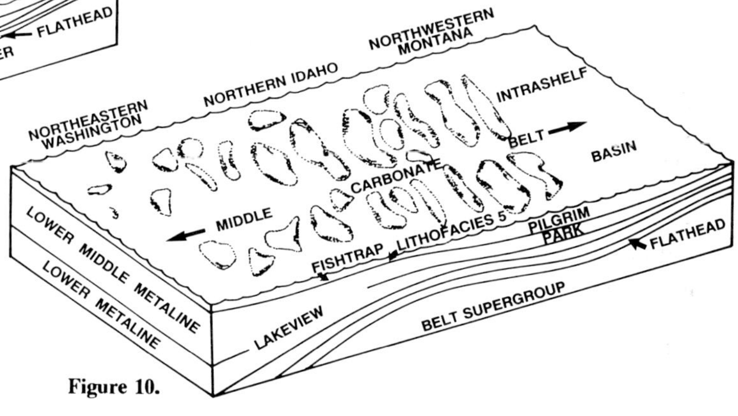

Figure 10.

Block diagrams illustrating the depositional setting.

Figure 7. After first major transgression into Montana during the early Middle Cambrian.
Figure 8. At the end of the first major upward-shallowing Grand Cycle.
Figure 9. During the clastic portion of the second Grand Cycle at the end of the Middle Camrian.
Figure 10. For the carbonate half of the second Grand Cycle during the early Late Cambrian.

3. Tidal range, current activity, and distribution of clastic facies in the intrashelf basin.
4. Vertical lithologic successions and depositional cycles on the shoal complex and in the intrashelf basin.
5. Thickness and distribution of distal carbonates on the outer edge of the shelf.

Near the end of the Middle Cambrian, renewed rapid transgression initiated another Grand Cycle with the deposition of the Park Shale, the Switchback Shale, and the Fishtrap lithofacies 4 over the shoal complex on the intrashelf side. This second cycle ended with carbonate deposition of the Pilgrim Formation in southwestern Montana, the Devil's Glen Dolomite in west-central Montana, lithofacies 5 of the Fishtrap Formation, the upper portion of the middle Metaline Formation in northeastern Washington, and possibly the top of the Lakeview Limestone (lithofacies 6) in northern Idaho. Figures 9 and 10 illustrate the clastic and carbonate stages of this second cycle. The algal-shoal complex was rebuilt but the location and nature of the seaward edge of the complex is difficult to determine. Aadland (1979, 1985) and Bush and Fischer (1981) suggested that the algal shoals did not extend over northern Idaho and northeastern Washington after their earlier development during Grand Cycle one. It is now believed that the upper Lakeview Limestone (lithofacies 6) and the upper middle Metaline represent this second Grand Cycle, where continuous deposition of subtidal to supratidal carbonates occurred far from the effects of clastics of the inner detrital belt. This interpretation implies that shallow water carbonates were deposited in northern Idaho and northeastern Washington from late Middle until early Late Cambrian.

The second Grand Cycle ended with renewed transgression that drowned the outer platform. The drowning destroyed the peritidal, algal-shoal complex in northeastern Washington and northern Idaho and prohibited its full development in western Montana. The record for this third cycle in southwestern Montana is represented by the Red Lion Formation (Hayden and Bush, 1987). The lower Dry Creek shale member of the Red Lion represents the lower clastic half-cycle. The overlying Sage Pebble conglomerate member represents the upper carbonate half-cycle. Regional trends of clastic content in the Red Lion and underlying Pilgrim Formation indicate that the Lemhi Arch, to the south in central Idaho, was active during this time. Increased clastic material from the Lemhi Arch and from the northeast contributed to the reduction in algal growth and shallow water carbonate production throughout this cycle.

The carbonate half of the last Cambrian cycle differs from the earlier two cycles. The Red Lion lacks the abundant mottled mudstones, oolitic grainstones and flat-lying cryptalgal bindstones of the Meagher, Pilgrim, and Fishtrap (lithofacies 1-5) of the earlier cycles. Upward shallowing in the Red Lion is indicated primarily by the presence of columnar algal growths that overlie subtidal, below normal wave depth limestones and siltstones. Thus, in west-central and southwestern Montana intertidal algal banks, instead of supratidal algal flats, developed in the intrashelf basin far from the cratonic shoreline during the shallowing at the end of the Cambrian. The very top of the Fishtrap in the Libby Trough contains cryptalgal bindstones representing the seaward side of the algal shoals, which had retreated from northwestern Washington.

The drowning of the outer platform in northeastern Washington is documented by below-wave-base, dark, sparsely fossiliferous limestones that overlie the peritidal dolostones of the middle Metaline Formation. The complex stratigraphic relations and the sparsity of fossil control between Cambrian and Ordovician units in northeastern Washington, as well as rapid north-south and east-west facies changes, prevent accurate determination of sequential events for the outer ramp. Deposition of upper Metaline distal carbonates occurred during the lower half-cycle at the same time as outer detrital belt deposition of lower Ledbetter equivalents seaward of Metaline Falls. During the upper carbonate half-cycle, transgression slowed and slope deposition increased at the ramp's edge, forming portions of the "Josephine Breccias" near Metaline Falls. J.A. Morton (unpublished manuscript, 1988) has recently documented that the "Josephine Breccias" and other isolated breccia bodies, which occur throughout the Middle Cambrian to Early Ordovician units in northeastern Washington, are debris and slump deposits. Figures 11 and 12 illustrate the depositional interpretation outlined herein for the last Cambrian Grand Cycle for northeastern Washington, northern Idaho, and northwestern Montana.

Although only one Early Ordovician outcrop has been identified in western Montana, regional depositional trends can be suggested. The isolated quartz arenite-dolostone sequence in the Libby Trough represents a remnant of an Early Ordovician Grand Cycle. Figure 13 illustrates a depositional model for the Early Ordovician that has algal banks over western Montana, clastics to the east, and outer detrital belt sediments (Ledbetter Formation) over northeastern Washington and northern Idaho. Slump and debris deposition continued during the Early Ordovican over northeastern Washington.

In summary, three Cambrian Grand Cycles outline deposition in northern Idaho and northwestern Montana during the Middle and Late Cambrian. Sedimentation was controlled by the migration and development of an extensive peritidal, algal-shoal complex that expanded during the carbonate half and contracted during the clastic half of each Grand Cycle.

ROAD LOG

This guide describes two days of field trips to examine Cambrian strata in northern Idaho and northwestern Montana. The first two stops near Bayview, Idaho, examine the Lakeview Limestone along Pend Oreille Lake. Three stops are between Bayview, Idaho, and Libby, Montana. The first stop is at a roadcut in the Lakeview

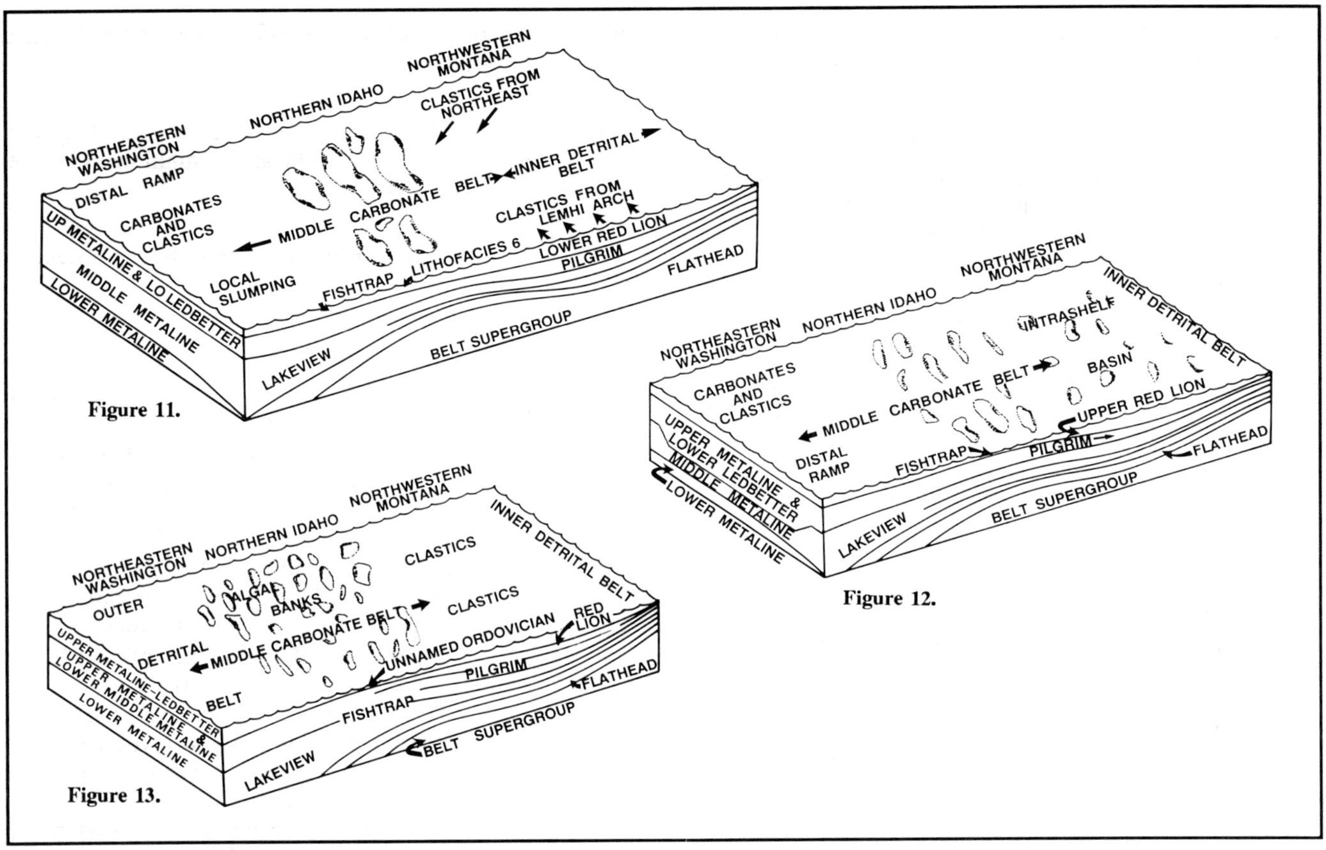

Block diagrams illustrating the depositional setting.

Figure 11. During the clastic portion of the third Grand Cycle during the Late Cambrian.
Figure 12. For thke carbonate half of the third Grand Cycle at the end of the Cambrian.
Figure 13. For the Early Ordovician Grand Cycle.

Limestone near Heron, Montana, and the second is at an isolated outcrop on Lime Butte south of Troy, Montana. The third stop is north of Moyie Springs, Idaho. The Pend Oreille Lake and Heron stops show distal ramp argillaceous limestones. The Lime Butte and Moyie Springs stops reveal overlying peritidal dolostones.

The second day, several stops are within the Libby Trough from Libby southeastward to Thompson Falls. These stops examine the stratigraphic sequence, from the early Middle Cambrian Flathead Sandstone to an Early Ordovician dolostone.

Day One: Spokane to Libby

The field trip for Day 1 contains five stops. Stop 1 is at Bayview, Idaho, to inspect the Cambrian outcrops along the shore of Pend Oreille Lake. Stops 2 and 3 are in lithofacies 1 of the Lakeview Formation. Stops 4 and 5 are in overlying shallow subtidal to supratidal dolostones. The outcrops are small and widely separated by exposures of Belt Supergroup rocks, but they represent a nearly complete stratigraphic section of upward-shallowing Middle Cambrian environments.

Mileage	Description
0.0	Begin mileage at Exit 281 in downtown Spokane, Washington. Head east toward Coeur d'Alene, Idaho, on Interstate 90.
10.5	Exit 291. Take exit ramp to stop sign. Turn left on Sullivan Road.
12.5	Junction with Washington Highway 53. Turn right towards Rathdrum, Idaho.
34.3	Junction with U.S. Highway 95. Turn north on U.S. 95. Proceed to Athol, Idaho.
44.3	Junction with Idaho Highway 54 to Bayview. Turn right to Bayview.

51.8 City limits of Bayview. Make a sharp right turn on unmarked paved road.

53.2 **Stop 1: Lakeview Limestone overview.**

Park on gravel pulloff on left. From this viewpoint, several Cambrian outcrops are visible in the distance along the shore of Pend Oreille Lake. Look to the northeast (left, as you face the lake) across the lake to the large light gray cliff with a roadcut transecting it. The cliffs are metamorphosed Lakeview Limestone units that are in fault contact with overlying Gold Creek Quartzite. The top of the ridge consists of Belt Supergroup rocks. West of the roadcut are granodiorite exposures of the Kaniksu batholith.

Look east across the lake to the resort village of Lakeview that is in a steep valley along the lake shore. The gray cliffs are Lakeview Limestone, and the prominent brown cliffs to the southeast are Gold Creek Quartzite resting on the Wallace Formation of the Belt Supergroup. One mile east of Lakeview along the shoreline are light-colored fresh exposures caused by landslides and quarrying in the upper Lakeview. Return on paved road toward Bayview.

53.7 **Stop 2: Lakeview Limestone (boat access only).**

Boats are necessary to reach Stop 2, which is an old cement plant loading area about 1.4 miles east of Lakeview (SW1/4, SW1/4, sec. 26, T. 54 N., R. 1 W.). This boat trip is 5 miles, and boaters should be aware of lake conditions. The exposures along the beach, directly below the old cement pillars, are in black lime mudstones of lithofacies 1 of the Lakeview Limestone. The exposures to the east of the pillars (left as you face the shore) are in the lowermost part of lithofacies 1.

Trilobites can be collected at several localities along the beach. *Elrathina* specimens are common on bedding planes in the lowermost units. The black lime mudstones at the top of the pillars contain numerous *Peronopsis* specimens.

This locality is believed to be the cement mine collecting site of Resser (1938). Motzer (1980) identified several specimens from the black lime mudstones along the shoreline and provided the following taxonomic list:

Brachipods; *Acrothele* sp., *Acrotreta* sp., *Linqulella* sp., Trilobites; *Peronopsis vonnerensis* (Resser), *Alokistocare* sp., *Clauaspidella* sp., *Oryctocephalus* sp., *Pagetia* sp., and *Zacanthoides* sp.

After Stop 2 is completed return to U.S. 95 at Athol.

62.1 Junction with U.S. 95. Turn north toward Sandpoint.

86.6 Junction in Sandpoint with Idaho State Highway (S.H.) 200. Turn right and remain on S.H. 200 toward Clark Fork.

111.4 Town of Clark Fork. Continue through town on S.H. 200.

124.8 **Stop 3: Heron outcrop.**

Park on small graveled area on the south side of highway. The Cambrian rocks at this stop are in the downfaulted block of the Hope fault. Although the rocks are faulted and poorly exposed, the stratigraphic sequence can be determined and correlations to the Lakeview Limestone can be made. The exposures in the roadcuts along the north side of the highway correlate to the lower portions of Lakeview lithofacies 1 and comprise 82 meters of laminated, sparsely fossiliferous, argillaceous lime mudstone.

The rocks between the highway and Clark Fork River on the south side are younger than the northern exposure. The southern outcrops consist of about 20 meters of massive, crystalline, gray weathering limestone that sandwiches a laminated, argillaceous, lime mudstone. The laminated unit contains concretions of black, fetid, pyritic, fossiliferous lime mudstone. The concretions are lenticular to ellipsoid and vary in size from 20 centimeters wide and 5 centimeters thick to 1.5 meters wide and 1 meter thick. Local collectors refer to these concretions as "sea biscuits" or "miner's biscuits" (Figure 14). Select surfaces within the concretations contain a trilobite-rich "hash" (Figure 15) first reported by Campbell and others (1937). This interval correlates to the black lime mudstones at the cement plant and to concretions noted 150 meters above the base of the Lakeview Formation (Figure 16). After Stop 3 continue east on S.H. 200.

129.5 Junction with U.S. Highway 56. Turn (north) left toward Libby on U.S. 56.

158.1 Milepost 29.

158.6 Turn left onto unmarked paved road.

160.6 Y-junction with another unmarked paved road. Make a sharp left turn.

161.8 Y-junction. Bear right on Keeler-Rattle Road.

Figure 14. Concretions from the lower Lakeview at Heron, Montana. Subdivisions on the scale equal 10 centimeters.

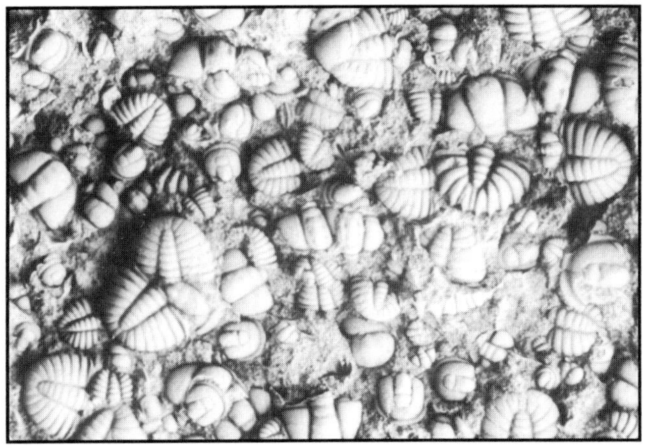

Figure 15. Example of trilobite "hash" collected from a concretion at the Heron, Montana, locality. Scale in centimeters.

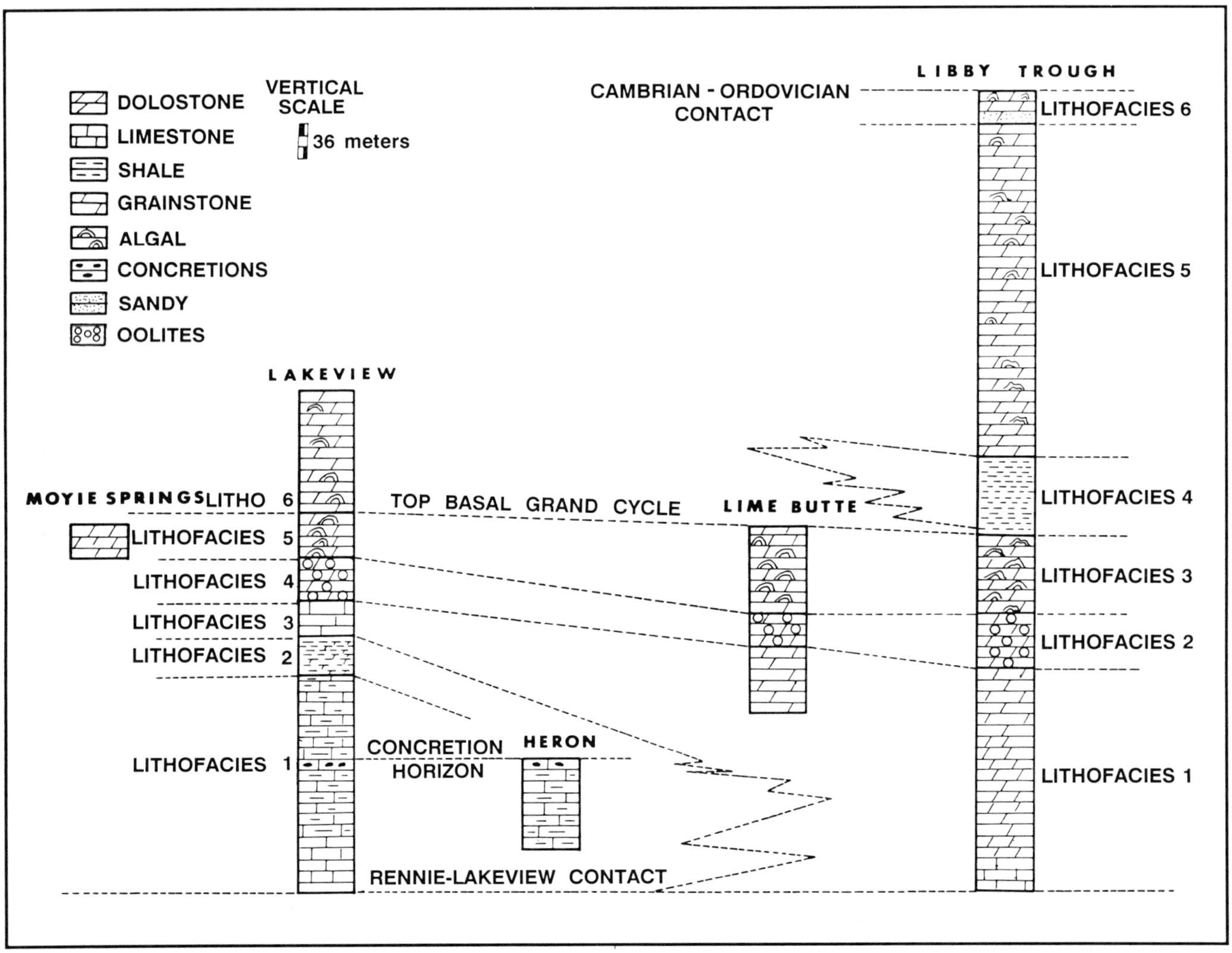

Figure 16. Stratigraphic cross-section showing correlations of isolated outcrops at Heron and Lime Butte, Montana, and Moyie Springs, Idaho, to the Lakeview and Fishtrap Formations.

162.9 Y-junction. Turn right and remain on Keeler-Rattle Road.

163.2 Unmarked junction. Turn right onto an unimproved gravel road.

164.0 Unmarked junction. Turn right onto a grass-covered logging road.

164.3 **Stop 4: Lime Butte outcrop.**

Approximately 170 meters of Cambrian carbonates crop out and form two buttes south of the road. The units strike northwest and dip southwest. A 500-meter traverse in a southwestward direction across strike goes up section. Of the two buttes, the southernmost is Lime Butte. J. E. Harrison (personal communication, 1988) has mapped these rocks as Cambrian in age; they are in fault contact with the Prichard Formation of the Belt Supergroup on the east plate of an off-shoot of the Moyie thrust. The sequence consists of dolomitized units ranging from brown and dark gray, mottled, partly exposed basal mudstones to upper, light gray, well-exposed, flat-lying, cryptalgal dolobindstones and oncolitic dolopackstones. The basal units contain a variety of mottled structures representing partial to complete bioturbation of the sediments. Minor intraclastic grainstones and flat-pebble conglomerates are interbedded with the mudstones. The upper cryptalgal dolobindstones and oncolitic dolopackstones are resistant and form cliffs above the basal mudstones. Cryptalgal bindstones dominate the upper units in beds 5 to 40 centimeters thick with minor oncolitic dolopackstones. Interbedded units (5-15 cm) of "birdseye" fenestrae, and structureless, light gray dolomudstones are also present.

Although no fossils have been identified, the sequence at Lime Butte can be correlated to the first upward-shallowing carbonate half of a Grand Cycle at other localities in northwestern Montana and is considered Middle Cambrian in age. The sequence of subtidal, mottled lime mudstones to cryptalgal dolobindstones is similar to lithofacies 1-3 of the Fishtrap Formation in the Libby Trough area to the southeast and to lithofacies 3-5 of the Lakeview Limestone to the southwest. The sequence at Lime Butte provides additional documentation of the development of the large Middle Cambrian peritidal algal-shoal complex which covered northern Idaho, northeastern Washington, and northwestern Montana.

After this stop return to U.S. 56.

170.0 Junction with U.S. 56. Turn left toward Libby.

175.3 Junction with U.S. Highway 2. Turn left toward Troy.

176.9 Entering Troy. Continue on U.S. 2.

191.2 Junction with old U.S. 2. Turn right.

203.6 Junction with gravel road. Make a sharp right onto the Deer Creek-Canuck Road.

204.6 Y-junction. Turn left onto Forest Service Road 435 toward Canuck Pass.

206.6 Unmarked junction. Turn left onto unimproved dirt road.

206.7 Y-junction. Bear left beyond metal Forest Service gate.

207.5 Grass-covered Y-junction. Bear left.

208.1 **Stop 5: Moyie Spring outcrop.**

The small roadcuts on the left side of the road have been mapped as Cambrian(?) on the east plate of the Moyie-Leonia fault (Burmester, 1985). The sequence consists primarily of alternating light gray and black, finely crystalline dolostone.

The light gray units (10 cm to 8 m) are plane-parallel bedded in places and are generally structureless with "birdseye" features visible locally. The black units, 10 centimeters to 3 meters thick, are finely crystalline dolomudstones with distinctive "birdseye" and other irregular "spots" of coarse white dolomite. Thin intraclastic dolowackestones and brecciated dolomudstones and oncolitic dolopackstones are also present.

The light gray and black interlayering with the distinctive "birdseye" texture in finely crystalline dolostone is characteristic of the middle Metaline Formation in northeastern Washington.

The lower portions of the middle Metaline are Middle Cambrian in age. However, the age relations of Metaline units have not been clarified, and the top of the middle Metaline could be Late Cambrian in age. Tentatively, the Moyie Springs dolostones are considered to have been part of the first upward-shallowing Middle Cambrian Grand Cycle. The sequence is correlated to the lower middle Metaline Formation, to the upper portion of the carbonates on

Lime Butte, and to lithofacies 3 of the Fishtrap Formation.

Return to U.S. 2. Turn left (east) and proceed to Libby. The road log for Day 2 starts in Libby.

Day Two: Libby to Thompson Falls

The field trip for Day 2 traverses Cambrian rocks exposed in three separate blocks of the Libby Trough. The travel route crosses all three blocks, but the stops are in the southernmost block where a complete section is present and numerous logging roads provide access.

Mileage	Description
0.0	Begin in Libby at junction of U.S. Highways 37 and 2. Proceed east on U.S. 2.
12.3	Bridge over Libby Creek.
13.0	Gray outcrops of Fishtrap Formation on right side of highway. The highway traverses the northernmost block of the Libby Trough. Exposures of Fishtrap Formation occur on both sides of the highway for the next 3 miles. Gale (1934) first described these rocks and suggested that the beds are overturned. Aadland (1979) agrees that the units are overturned and provides details for 246 meters of the Fishtrap Formation. Much of the section is fractured and faulted; consequently, correlation to the other two blocks is difficult. The Flathead Sandstone in the northern block may be abnormally thin since it occurs only as minor float, one cobble thick, between exposures of the Belt Supergroup and the Gordon Shale. A logging road cut (SW1⁄4, SW1⁄4, sec. 14, T. 28 N., R. 30 W.) exposes 26 meters of maroon and green Gordon Shale overlain by 20 meters of black, mottled lime mudstone which is in turn overlain by fractured, crystalline, brown dolostone. The Flathead Sandstone, Gordon Shale, and basal lime mudstone unit of the Fishtrap Formation are all thinner than their equivalents in the southernmost block, indicating a northward thinning.
13.7	Bridge over Swamp Creek
15.2	A small exposure on the right contains Gordon float with trilobite "hash" on bedding planes. This location is close to a site reported in Gale (1934) of a railroad cut that has since been overgrown. The best exposure in the area occurs along the logging road referred to earlier. Gale (1934) concluded that the shale is Middle Cambrian in age and that the fauna have similar characteristics to those described by Walcott (1917) from the Gordon in Powell County 60 kilometers to the east. Gale (1934) also noted a resemblance to fauna reported in the Spence Shale of southern Idaho. Of particular note are the extremely well-preserved large *Hyolithes* specimens that reach 75 centimeters in length.
15.8	Bridge over Swamp Creek.
16.1	Vertical outcrops on both sides of the highway are in the Belt Supergroup.
24.6	Bridge over Fischer River. The highway for the next 14 miles passes south of the middle block of the Libby Trough. Only a few gray outcrops of the Fishtrap Formation are visible from the highway. Johns (1970) mapped the Cambrian rocks in the middle exposure but did not differentiate individual units. Aadland (1979) reported a composite section of 490 meters of Fishtrap from this area.
31.5	Enter the Houghton Creek fire region.
38.5	Bridge over Fischer River.
47.7	Enter Flathead County.
48.6	Turn right onto Thompson River Road.
Reset mileage	
0.0	Thompson River Road turnoff.
6.5	Bridge across the headwaters of Thompson River. Turn left onto private logging road and continue south along the river.
11.0	Junction at Bend Campground. Take sharp right and bear left on Bend-Vermilion Road (U.S. Forest Service No. 7593).
11.3	Y-junction. Bear left to Fishtrap Lake.
12.3	Y-junction. Bear left on Bend-Vermilion Road.
18.8	T-junction with Fishtrap Creek Road. Turn right.
18.9	Junction with Bend-Vermilion Road. Turn left to Fishtrap Lake.
19.2	Junction, unmarked. Bear right.

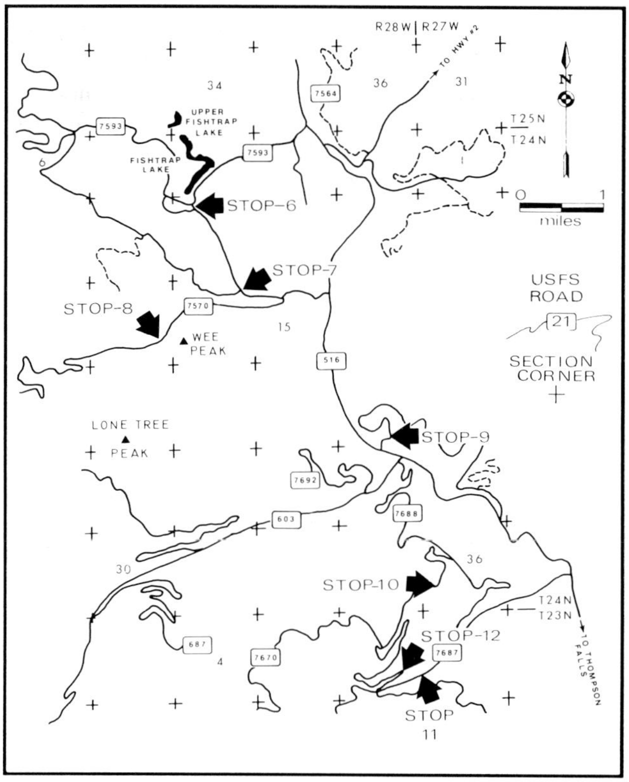

Figure 17. Location map showing Stops 6-12 for the second day.

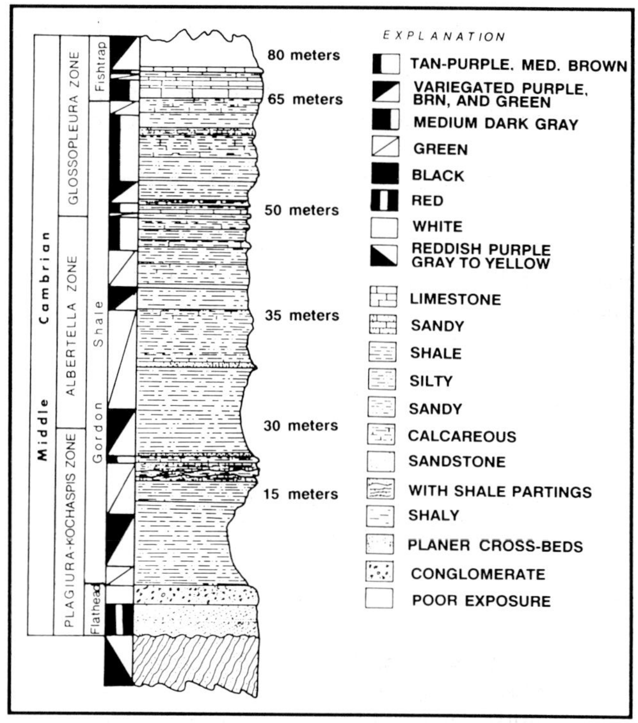

Figure 18. Stratigraphic section of the Gordon Shale and Fishtrap Formation at Stop 7. Modified from O'Malley (1985).

20.2 Fishtrap Lake on right (Figure 17).

21.8 **Stop 6: Flathead Sandstone.**

Fishtrap Campground on right. Park vehicles and walk back across bridge on the main road. Turn right on unused logging road just past the bridge and walk about 100 meters to the exposure of Flathead Sandstone over Libby Formation. At this locality a "lag gravel" occurs at the Precambrian-Cambrian contact. Note that the gravels are one layer thick and consist primarily of "Belt siltites" that are well rounded and generally bladed in shape and range from 1 to 5 centimeters in diameter.

The lower unit of the Flathead Sandstone is easily accessible at this locality, but the upper conglomeratic unit is only exposed at the eastern end of the outcrop. The lower maroon unit is considered continental and the upper part marine. In the Libby Trough, it can be shown that the lower unit laterally grades into beds that are lighter in color and are only locally maroon. A few centimeters above the Precambrian rocks, these light-colored beds contain burrows, worm markings, and "arthropod" trails; therefore, all of the Flathead Sandstone at this locality is considered Cambrian.

21.1 Junction. Turn left on Radio Creek Road.

21.8 **Stop 7: Gordon Shale.**

Outcrops in the upper Gordon Shale along right side of road contain brachiopods and trilobites at several locations. Walk ahead toward junction with Forest Service Road No. 7553. A nearly complete section of Gordon Shale is exposed in cuts below the road. O'Malley's (1985) stratigraphic section is presented on Figure 18. The gradational contact with the overlying Fishtrap is well exposed. Mottled, black lime mudstones are interbedded with the shales. Turn right and walk up Forest Service Road No. 7553 to examine the basal portions of lithofacies 1 of the Fishtrap Formation. The lime mudstones in the Libby Trough grade upward into light-colored mottled dolomudstones with occasional interbedded intraclastic dolowackestones and dolograinstones.

23.0 Junction with Forest Service Road No. 7553. Continue straight ahead.

23.6 Junction with Forest Service Road No. 7570 (Beartrap Road). Make a sharp right turn onto Beartrap Road.

23.8 Junction with Forest Service Road No. 7692. Continue straight ahead on Beartrap Road.

24.8 Junction with Wee Peak Road. Continue straight ahead.

25.3 Cliffs on right, across the gulch, are quartzite which overlie the Fishtrap Formation.

25.9 **Stop 8: Unnamed Ordovician.**

Park along road and walk northwest about 300 meters through the brush across Beartrap Creek to the base of the outcrop. The basal crystalline gray dolostone is the top of the Fishtrap Formation (lithofacies 6). Overlying the dolostone is a 65-meter-thick quartz arenite, which is overlain by 37 meters of grayish red dolostone. The upper dolostones contain Early Ordovician conodonts and represent the only known Ordovician in western Montana. The quartzites have been mapped at several localities on the Fishtrap quadrangle (R. M. Hague, unpublished map). No conodonts were obtained from the underlying Fishtrap Formation. The quartzite to dolostone sequence represents a portion of an Early Ordovician Grand Cycle.

Lithofacies 6 is underlain by an upward-shallowing cycle of the Fishtrap lithofacies 4 and 5. Therefore lithofacies 6, although thin, is interpreted to represent one Grand Cycle in Late Cambrian time correlative with the Red Lion Formation of southwestern Montana.

28.2 Junction with Radio Creek Road. Bear right.

29.0 Junction with Fishtrap Road (No. 516). Turn right.

31.0 Junction with Shale Ridge Road (No. 7691). Turn left.

31.4 **Stop 9: Flathead Sandstone and Gordon Shale.**

Switchback to left. Park and walk along old logging road that cuts off from the switchback and heads southeast. Walk about 60 meters to the end of old logging road to outcrops of maroon-buff Flathead Sandstone. As you return to vehicle, you walk up the sequence. The lower Flathead with Maroon-buff quartz arenite consists of parallel laminations and low angle cross-beds. The upper Flathead consists of conglomeratic quartz arenite with numerous vertical worm burrows. The lower Gordon Shale consists of green shales with glauconitic sandstone interbeds. Horizontal burrows and worm trails are common on bedding planes. Green "cornflake-like" shale exposed in the switchback of the main road contains occasional trilobite and brachiopod fragments.

31.8 Return to Fishtrap Road. Turn left.

32.2 Junction with U.S. Forest Service Road 603, (West Fork Fishtrap Creek Road). Take sharp switchback to right.

32.3 Cross over Fishtrap Creek.

32.4 Junction. Turn left and continue left on Fishtrap-Beatrice Saddle Road.

32.6 Quarry on left in Flathead exposing both lower and upper units.

33.1 Road splits. Bear right.

35.9 Junction. Take sharp right on Upper Beatrice Creek Road.

36.3 Y-junction. Bear left.

37.1 **Stop 10: Fishtrap Formation (lithofacies 2).**

Outcrops occur on both sides of road in sharp curve. Light gray outcrops along both sides of the road represent about 6 meters of the oolitic and peloidal dolograinstones of the Fishtrap Formation (lithofacies 2). At this locality, faintly cross-bedded oolitic dolograinstones are interbedded with mottled dolomudstones and flat-lying cryptalgal dolobindstones. Aadland (1979, 1985) correlates the rocks of lithofacies 2 with the upper portion of the Pagoda Limestone in central Montana and suggests a possible correlation to parts of the Meagher Formation of southwestern Montana. Westward, this unit correlates with lithofacies 4 of the Lakeview Formation. After examination of the outcrop, return to Lower Beatrice Road.

39.2 Y-junction. Bear right on Beatrice Creek Road.

39.4 Junction. Continue straight ahead.

40.3 Sharp switchback and junction with Lower Beatrice Road. Turn right.

42.0 **Stop 11: Fishtrap Formation (lithofacies 3).**

Outcrops next to road are near the top of lithofacies 3 of the Fishtrap Formation. They are characterized by flat-lying cryptalgal dolobindstones and crystalline fine-grained dolomudstones. The laminae vary from crinkled and crenulated to wavey and continuous. Small-scale disconformities such as erosional surfaces and local brecciation are common. Laminoid fenestral fabric is also common. Aadland (1979, 1985) correlates lithofacies 3 eastward to the Steamboat Limestone. Westward, it correlates to the carbonate sequence on Lime Butte and to lithofacies 5 of the Lakeview Formation and to the lower portion of the middle Metaline Formation. Fishtrap lithofacies 1-3 represent the carbonate half of the first major upward-shallowing Grand Cycle, and lithofacies 3 was part of the first major peritidal, algal-shoal complex that developed over northern Idaho, northeastern Washington, and western Montana.

42.6 Take sharp right on unmarked road.

43.1 **Stop 12: Fishtrap Formation (lithofacies 4 and 5).**

Switchback to left. White outcrops exposed in roadcuts belong to Fishtrap lithofacies 5. Much of lithofacies 5 is recrystallized dolostone. Where features are identifiable, flat-lying cryptalgal dolobindstones are most common. Minor dolomudstones and peloidal dolopackstones are also present. Walk about 700 meters down the logging road that veers to the right from bend in switchback. The road begins in flat-lying cryptalgal dolobindstones of lithofacies 5, crosses lithofacies 4, and ends in lithofacies 3.

This locality is one of the few places that the clastics of lithofacies 4 can be examined. The lithofacies consists predominately of shale interbedded with argillaceous dolostones. The shales are laminated and interbedded with occasional siltstones. Some laminae are continuous, while others are discontinuous and lens shaped. Most of the laminae pinch and swell. Within a few bands, parallel laminae are noted, as well as low angle cross-laminae. Occasional elongate flat-pebble shale clasts are also present. Lithofacies 4 and 5 represents the second Grand Cycle in the Fishtrap Formation. Lithofacies 4 is the clastic basal half. Aadland (1979) correlated it with the Switchback Shale (late Middle to early Late Cambrian) of northwestern Montana and the Park Shale (late Middle to early Late Cambrian) of southwestern Montana. The interpretation from Grand Cycle correlation is that lithofacies 4 and 5 range from late Middle to mid-Late Cambrian in age.

In summary, the Fishtrap Formation can be subdivided into six major lithofacies that represent three upward-shallowing Grand Cycles which formed during the contraction and expansion of an extensive peritidal algal-shoal complex. Ordovician strata represent a fourth Grand Cycle. The cycles are useful for correlations from outcrop to outcrop, for regional correlations, and for interpreting depositional environments. Stop 12 is the last one for Day Two.

43.6 Return to Lower Beatrice Creek Road and turn left.

44.7 Junction with Upper Beatrice Creek Road. Continue straight ahead downhill.

46.7 Junction with Fishtrap Creek Road. Take a sharp right.

53.3 Junction with Thompson River Road. Turn right.

53.9 Junction at bridge with private logging road. Continue straight ahead.

19.7 Junction with S.H. 200. Turn right. The return trip to Spokane via Sandpoint is 170 miles.

ACKNOWLEDGEMENTS

The manuscript benefited from reviews by Peter E. Isaacson and Larry Davis. W. Patrick Seward assisted with the initial editing, and David C. Kachek assisted with the early phases of the field work. Mark and Shirley McFaddan field checked mileage and directional instructions. Special thanks to Linda L. Hayden who assisted with the field work and provided the encouragement to finish the project.

REFERENCES

Aadland, R. K., 1979, The Cambrian stratigraphy of the Libby Trough, northwestern Montana: University of Idaho Ph.D. thesis, 236 p.

———, 1985, Lithofacies of the Middle-Upper Cambrian sequence, Libby syncline, Montana: Montana Bureau of Mines and Geology Open-File Report 153, 52 p.

Aitken, J. D., 1966, Middle Cambrian to Middle Ordovician cyclic sedimentation, southern Rocky Mountains of Alberta: Bulletin of Canadian Petroleum Geology, v. 14, p. 405-411.

———, 1978, Revised model for depositional Grand Cycles, Cambrian of the southern Rocky Mountains, Canada: Bulletin of Canadian Petroleum Geology, v. 26, no. 4, p. 515-542.

———, 1981, Generalizations about Grand Cycles, *in* M. E. Taylor, editor, Short Papers for the Second International Symposium on the Cambrian System: U.S. Geological Survey Open-File Report 81-743, p. 8-14.

Apgar, J. L., 1986, Stratigraphy, sedimentology and diagenesis of the Flathead Sandstone, Libby Trough, Montana: University of Idaho M.S. thesis, 95 p.

Burmester, R. F., 1985, Preliminary geologic map of the Eastport area, Idaho, Montana: U.S. Geological Survey Open-File Map 85-0517.

Bush, J. H., 1977, Lithologic units of the Lakeview Limestone, Bonner County, Idaho: Geological Society of America, Abstracts with Programs, Rocky Mountain Section, v. 9, no. 6, p. 712-713.

Bush, J. H., H. J. Fischer, and R. K. Aadland, 1980, The importance of a Middle and Upper Cambrian algal shoal-tidal barrier over northern Idaho: Geological Society of America, Abstracts with Programs, Cordilleran Section, v. 12, no. 3, p. 100.

Bush, J. H., and H. J. Fischer, 1981, Stratigraphic and depositional summary for Middle and Upper Cambrian strata in northwestern Montana, northeastern Washington, and northern Idaho, *in* M. E. Taylor, editor, Short Papers for the Second International Symposium on the Cambrian System: U.S. Geological Survey Open-File Report, 81-743, p. 42-46.

Bush, J. H., D. C. Kachek, and G. D. Webster, 1985, Newly discovered Ordovician strata, Libby Trough, northwestern Montana: Geological Society of America, Abstracts with Programs, v. 17, no. 4, p. 211.

Bush, J. H., and L. L. Hayden, 1987, Middle and Upper Cambrian platform evolution and paleogeography, northwestern Montana, northern Idaho and northeastern Washington: Association American Petroleum Geologists, Abstracts with Programs, v. 71, no. 8, p. 1002.

Campbell, Ian, R. P. Sharp, and H. R. Gale, 1937, A new locality for Middle Cambrian fossils near Noxon, Montana: American Journal of Science, 5th series, v. 34, no. 204, p. 411-421.

Chow, Nancy, and N. P. James, 1987, Cambrian Grand Cycles: A northern Appalachian perspective: Geological Society of America Bulletin, v. 98, no. 4, p. 418-429.

Fritz, W. H., 1975, Broad correlations of some Lower and Middle Cambrian strata in North American Cordillera: Geological Survey of Canada Paper 75-1, part A, p. 533-540.

Gale, H. R., 1934, Note on Cambrian fossils near Libby, northwestern Montana: Journal of Geology, v. 42, p. 174-179.

Hayden, L. L., and J. H. Bush, 1987, Stratigraphy and depositional environment of the Upper Cambrian Red Lion Formation, southwestern Montana: Association of American Petroleum Geologists, Abstracts with Programs, v. 71, no. 8, p. 1007.

———, 1988, Description and significance of three isolated Cambrian outcrops in northern Idaho and northwestern Montana: Geological Society of America, Abstracts with Programs, Rocky Mountain Section, v. 20, no. 6, p. 420.

Harrison, J. E., and D. A. Jobin, 1965, Geologic map of the Packsaddle Mountain quadrangle, Idaho: U.S. Geological Survey Quadrangle Map GQ-375, with 4 p. text.

Healy, M. P., 1986, Sedimentology and lithofacies of the Pilgrim Formation (Upper Cambrian) west-central Montana: University of Idaho M. S. thesis, 146 p.

Holland, L. H., editor, 1971, Cambrian of the New World: Wiley-Interscience, 456 p.

Johns, W. M., 1970, Geology and mineral deposits of Lincoln and Flathead counties, Montana: Montana Bureau of Mines and Geology Bulletin 79, 182 p.

Keim, J. W., and R. J. Rector, 1964, Paleozoic rocks in northwestern Montana: A newly recognized occurrence: Geological Society of America Bulletin, v. 75, p. 575-578.

Lochman-Balk, Christina, 1971, The Cambrian of the craton of the United States, *in* C. H. Holland, editor, Cambrian of the New World: Wiley-Interscience, p. 79-167.

———, 1972, Cambrian System, *in* W. W. Mallory, editor, Geologic Atlas of the Rocky Mountain Region:

Rocky Mountain Association Geologists, Hirshfeld Press, Denver, p. 60-75.

Martin, W. D., H. J. Fischer, R. J. Keogh, and K. Moore, 1980, The petrology of the limestone formations (Middle and Upper Cambrian) northwestern Wind River Basin, Wyoming: Wyoming Geological Association, Thirty-second Annual Field Conference Guidebook, p. 37-52.

Motzer, M. E., 1980, Paleoenvironments of the Lower Lakeview Limestone (Middle Cambrian), Bonner County, Idaho: University of Idaho M.S. thesis, 95 p.

Mount, J. F., and S. M. Rowland, 1981, Grand Cycle A (Lower Cambrian) of the southern Great Basin: A product of differential rates of relative sea-level rise, *in* M.E. Taylor, editor, Short Papers for the 2nd International Symposium on the Cambrian System: U.S. Geological Survey Open-File Report 81-743, p. 143-146.

O'Malley, D. P., 1985, Trilobite biostratigraphy of the Gordon Shale in the southern part of the Libby Trough, Sanders County, Montana: Washington State University M.S. thesis, 144 p.

Oriel, S. S., and F. C. Armstrong, 1971, Uppermost Precambrian and lowest Cambrian rocks in southeastern Idaho: U.S. Geological Survey Professional Paper 394, 52 p.

Palmer, A. R., 1960, Some aspects of the early Upper Cambrian stratigraphy of White Pine County, Nevada and vicinity: Intermountain Association of Petroleum Geologists Eleventh Annual Field Conference, p. 53-58.

———, 1971, The Cambrian of the Great Basin and adjacent areas, western U.S., *in* C. H. Holland, editor, Lower Paleozoic Rocks of the World, v. 1, Cambrian of the New World: Wiley-Interscience, p. 1-78.

———, 1981, On the correlatability of Grand Cycle tops, *in* M. E. Taylor, editor, Short Papers for the 2nd International Symposium on the Cambrian System: U.S. Geological Survey Open File Report 81-743, p. 156-159.

Palmer, A. R., and R. B. Halley, 1979, Physical stratigraphy and trilobite biostratigraphy of the Carrara Formation (Lower and Middle Cambrian) in the southern Great Basin: U.S. Geological Survey Professional Paper 1047, 131 p.

Resser, C. E., 1938, Middle Cambrian fossils from Pend Oreille Lake, Idaho: Smithsonian Miscellaneous Collection, v. 97, no. 3, 12 p.

Reineck, H. E., 1972, Tidal flats, *in* J. K. Rigby and W. K. Hamblin, editors, Recognition of ancient sedimentary environments: Society of Economic Paleontologists and Mineralogists Special Publication 16, p. 146-159.

Robison, R. A., 1960, Lower and Middle Cambrian stratigraphy of the eastern Great Basin: Intermountain Association of Petroleum Geologists-Eastern Nevada Geological Society Guidebook, Joint Field Conference, p. 43-52.

———, 1964, Late Middle Cambrian faunas from western Utah: Journal of Paleontology v. 38, p. 510-566.

———, 1976, Middle Cambrian trilobite biostratigraphy of the Great Basin: Geology Studies Brigham Young University, v. 23, part 2, p. 93-109.

Sampson, Edward, 1928, Geology and silver ore deposits of the Pend Oreille district, Idaho: Idaho Bureau of Mines and Geology Pamphlet 31, 25 p.

Wanless, H. R., 1975, Carbonate tidal flats of the Grand Canyon Cambrian, *in* R. N. Ginsburg, editor, Tidal Deposits: Springer-Verlag, p. 269-277.

Walcott, C. D., 1917, The *Albertella* fauna in British Columbia and Montana: Smithsonian Miscellaneous Collection, v. 67, p. 9-59.

Chapter Four
Coeur d'Alene Mining District

Sacred Heart Mission at Cataldo, built in 1853, is the oldest standing building in Idaho. *Photograph courtesy of Idaho Division of Travel Promotion.*

The Geology and Alteration of the Gem Stocks, Shoshone County, Idaho

D. Kate Schalck [1]

INTRODUCTION

The Gem stocks are zoned, monzonitic to syenitic intrusions located northeast of Wallace, Idaho, in the Coeur d'Alene Mining District (Figure 1). These intrusions are subdivided into the south Gem stock and the north Gem stock (Figure 2). An associated group of intrusions, the Murray stocks, is located east of Murray, and the Dago Peak stocks are west of Dobson Pass (Parkinson, 1984). The Gem and Murray stocks exhibit a prominent northeast-southwest alignment, but the Dago Peak stocks have been offset from this trend by movement along the Dobson Pass fault. All these intrusions are small in areal extent, but it is known from mining that they get larger with depth (Hobbs and others, 1965). The largest, the south Gem stock, is exposed over 3 square miles (7.8 square kilometers). The exposed surface of each intrusion decreases to the northeast (Figure 2). All of the stocks were emplaced in the Precambrian Belt Super-

[1] *Washington State University, Pullman, Washington 99164.*

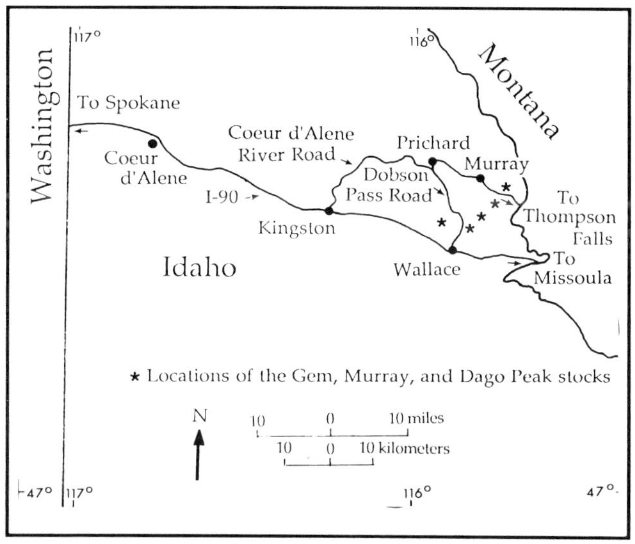

Figure 1. Location map showing the Gem, Murray, and Dago Peak stocks and the major roads for the field trip of the Gem stocks.

Schalck, D. K., 1989, The geology and alteration of the Gem stocks, Shoshone County, Idaho, *in* V. E. Chamberlain, R. M. Breckenridge, and Bill Bonnichsen, editors, Guidebook to the Geology of Northern and Western Idaho and Surrounding Area: Idaho Geological Survey Bulletin 28, p. 125-135.

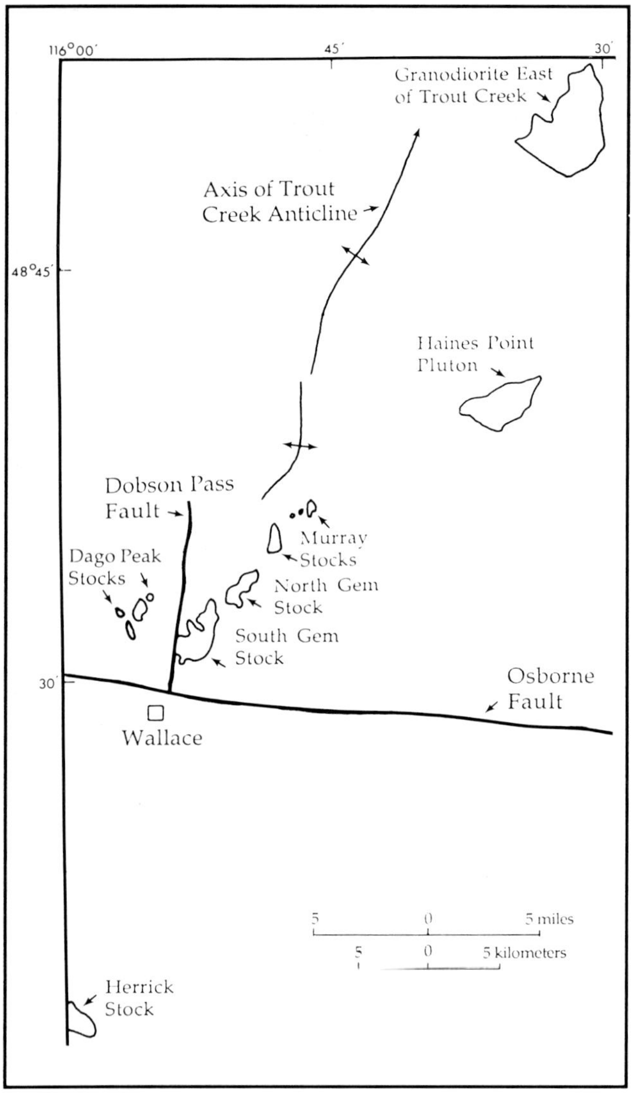

Figure 2. Locations of Cretaceous intrusions and major structures (Modified from Harrison and others, 1986).

group rocks. On a regional scale, the Gem stocks are part of a northeast-trending line of intrusions that include the Herrick stock to the southwest in the St. Joe River area. The Haines Point pluton and the granodiorite east of Trout Creek are northeast of the Gem stocks in Montana. According to Hobbs and others (1965), the intrusions north of Wallace roughly parallel the axis of the Trout Creek anticline (Figure 2).

IGNEOUS ROCK TYPES

Mafic Syenite

The mafic syenite (Streckeisen, 1975) is the most mafic unit and commonly contains a pyroxene that optically resembles augite. This unit includes a variety of rock types that are closely related in the field. These rock types include porphyritic mafic syenite with aligned potassium feldspar phenocrysts, equigranular mafic syenite, and mafic clots composed of hornblende, augite, and plagioclase. I interpret the alignment of the potassium feldspar crystals as primary flow foliation. Sphene and magnetite are generally visible in mafic syenite.

The mafic syenite is cut by aplitic hornblende syenite dikes and sparse quartz veins associated with silicification of the mafic syenite. Quartz may also be visible as an interstitial mineral between other crystals. The best exposures of mafic syenite are found in the southwest part of the south Gem stock (Figure 3). Small portions of the north Gem stock are also composed of mafic syenite, but these are too small to be shown on Figure 3.

Leucocratic Alkali-Feldspar Syenite

Leucocratic, alkali-feldspar syenite (Streckeisen, 1975) is very coarse grained and composed almost entirely of massive, gray potassium feldspar with up to 5 percent hornblende. This rock type is commonly stained with iron oxide and may contain disseminated sulfides. The unit grades into mafic syenite and monzonite, although at some locations float indicates that the leucocratic syenite has been brecciated by monzonite.

Monzonite

The monzonite (Streckeisen, 1975) can be divided into two phases: one is porphyritic and contains potassium feldspar megacrysts, and the other is equigranular. The potassium feldspar in the equigranular portion of the monzonite may be pink or white. Both phases may at some locations contain enough quartz to be quartz monzonites. This rock type everywhere contains visible hornblende and plagioclase in hand specimen. Accessory biotite is also present. The two phases of monzonite grade into each other. The monzonite with the pink potassium feldspar crystals and megacrysts is confined to the border zones of the intrusions. The north and south Gem stocks are composed predominately of equigranular monzonite (Figure 3).

Quartz Alkali-Feldspar Syenite

Quartz, alkali-feldspar syenite (Streckeisen, 1975), which is porphyritic, is characterized by the presence of biotite, with or without muscovite, and by the absence of hornblende. Zoned and perthitic, pink potassium feldspar phenocrysts are set in a groundmass of albite (Parkinson, 1984). Quartz, alkali-feldspar syenite is almost everywhere cut by quartz veins. I regard the quartz,

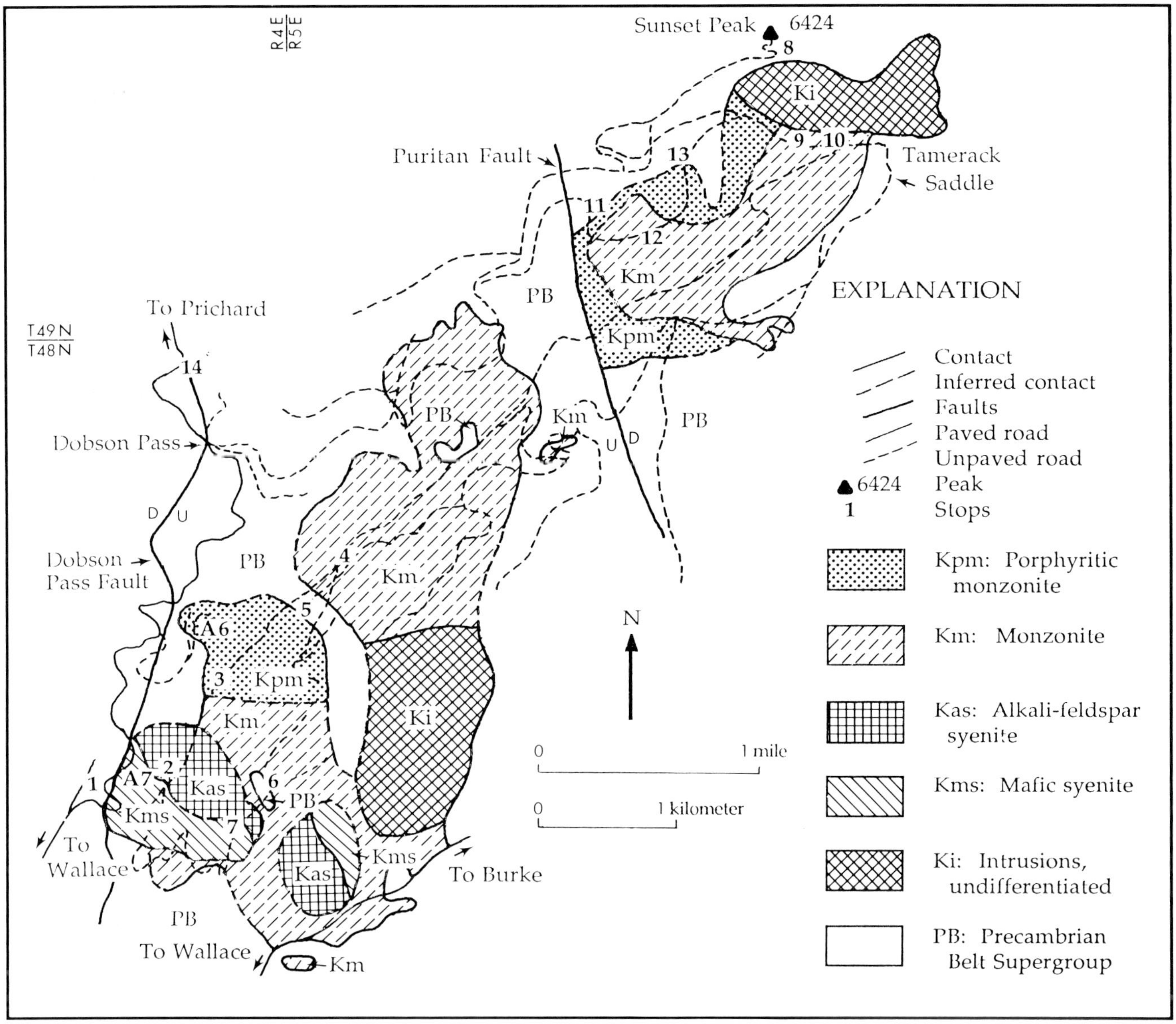

Figure 3. Simplified geologic map of the south and north Gem stocks (structures and Ki modified from Hobbs and others, 1965; remainder compiled from unpublished mapping by D. Kate Schalck).

alkali-feldspar syenite as the most differentiated portion of the intrusions. It is confined to the Round Top area near Murray.

ALTERATION AND MINERALIZATION

Zones of alteration are very common in the Gem stocks and include epidote-chlorite, argillic, and sericitic alteration and silicification. Epidote and chlorite alteration types are confined to phases that have significant percentages of mafic minerals. Portions of all of the stocks have been argillically altered and are characterized by clays in feldspars. Argillic alteration is also present around quartz veins in the Murray stocks. Sericitic alteration is confined to the monzonite and may be structurally controlled. Quartz veins and veinlets are present in all phases of the Gem stocks except the leucocratic, alkali-feldspar syenite.

Epidote veins, accompanied by epidote replacement of mafic minerals, are present in the phases of the stocks that contain significant amounts of mafic minerals. Chlorite alteration may be associated with the epidote alteration, and these two alteration types are most abundant in the mafic syenite. Epidote veins are also present in the monzonite.

Argillic alteration is sporadically present throughout the Gem, Dago Peak, and Murray stocks. The presence of clays within the feldspars is shown by the cloudy appearance of these crystals. Intense argillic alteration of the feldspars is present adjacent to the quartz veins in the Murray stocks.

Sericitic alteration is confined to the monzonite and in most places is fracture controlled. Oxides may be associated with sericite on fractures. Quartz veins and aplitic dikes are generally found within zones of pervasive sericitic alteration.

Quartz veins and veinlets, with or without associated silicification, are present in the mafic syenite, the monzonite, and the quartz, alkali-feldspar syenite. Other alteration types generally are associated with the quartz veins. Silicification and rare disseminated iron sulfides are found with quartz veins in the mafic syenite. Sericitic alteration of the monzonite is most intense where it is associated with quartz veinlets. Quartz veins occur as stockworks in the alkali feldspar syenite, and argillic alteration is present near the veins.

Iron sulfides are common in the monzonitic rocks found on the mine dumps in the area, but sulfides are not as common in the rocks found on the surface. The absence of sulfides in the rocks on the surface may be the result of weathering and oxidation processes that have removed the sulfides and left oxides behind. The alkali-feldspar syenite is an example of this process. This rock type contains disseminated iron sulfides at Stop 9, but is intensely iron oxide stained at Stop 2.

CONTACTS AND FAULTS

The contacts of the Gem stocks with the Belt rocks are irregular and complex and can only be seen in a few roadcuts and in the mines. The Gem stocks intrude the Prichard Formation on the northwest side and the Burke Formation on the southeast side (Harrison and others, 1986). Numerous inclusions of altered and unaltered Belt rocks occur in the intrusions and are particularly common near the upper contacts. Mafic inclusions also increase in abundance near the contacts with the Belt rocks. Hobbs and others (1965, p. 49) suggested that these mafic inclusions are partly assimilated Belt rocks, because they contain "granular aggregates of garnets and small remnants of undigested rock," and my study supports this interpretation. At Stop 12, some inclusions of Belt rocks are partially to totally digested. Several roof pendants of Belt rocks occur in the East Fork of Ninemile Creek (Figure 3).

Faults are present in the vicinity of the Gem stocks (Harrison and others, 1986) but may be difficult or impossible to pinpoint on the surface. My mapping confirms two of the major faults mapped in the area (Figure 2). The Dobson Pass fault is marked by a linear pattern of sheared and recrystallized pink potassium feldspar and iron oxides. The Puritan fault is present in the Interstate-Callahan Mine, but not well exposed on the surface (Hobbs and others, 1965). The southwest end of the north Gem stock is truncated by the Puritan fault (Figure 3). Other faults mapped by Harrison and others (1986) in this area have no surface expression.

AGE OF THE INTRUSIONS

The Gem stocks and associated intrusions yield conflicting dates, but a general pattern is present (Marvin and others, 1984). An emplacement age for the granodiorite east of Trout Creek, Montana, was established by Marvin and others (1984) with a concordia diagram of isotopic U-Pb data from zircon. The lower concordia intercept yields an emplacement age of 102±8Ma, and this age agrees with a concordant U-Th-Pb date on monazite of 102 to 106 Ma (Marvin and others, 1984). The U-Pb date on zircon at 103±8 Ma and the K-Ar date on hornblende at 99.4±3.4 Ma closely correspond to the other dates (Marvin and others, 1984). Therefore, the approximate 100 Ma age for the granodiorite east of Trout Creek seems to be valid.

The dates for the Haines Point pluton and the Gem stocks are not as numerous or as well established as the dates for the granodiorite east of Trout Creek, and they are more difficult to interpret. Micas from syenitic rocks from the Haines Point pluton yielded two K-Ar dates of 106±3 Ma and 107±4 Ma, but K-Ar from hornblende gives a date of 171±4 Ma (Marvin and others, 1984). Marvin and others (1984) suggest that the 107 Ma date corresponds to the age of fenitization of the Haines Point pluton, and that the older date either is the age of the intrusion or may be the result of excess argon in the hornblende.

The Gem stocks also yielded conflicting dates, which Marvin and others (1984) have summarized. Two Pb-alpha dates on zircon from granodiorite yielded 94±10 and 116±10 Ma (Larsen and others, 1958). These dates bracket the regional intrusive event of approximately 100 million years. McDowell (1971) dated hornblende from quartz monzonite in the Gem stocks by K-Ar and obtained two dates, 131±4 Ma and 137±4 Ma. Cretaceous dates have also been obtained from the Hercules Mine on the southeast side of the north Gem stock (Armstrong, 1975). Biotite associated with a vein from this mine yielded two K-Ar dates of 122±6 Ma and 104±5 Ma, but dates on lead obtained from the Hercules Mine are Precambrian (Long and others, 1960; Cannon and others, 1962).

In summary, the dates of the group of intrusions suggest an emplacement age of about 100 Ma (Marvin and others, 1984). This age is older than the Idaho batholith, which ranges from 95 to 70 Ma (Lewis and others, 1987).

It should be noted that the hornblende from both the Gem stocks and the Haines Point pluton yields K-Ar dates that appear to be anomalously old (Marvin and others, 1984). These may be the result of excess argon or may represent the actual ages of these intrusions.

GENESIS OF THE GEM, MURRAY, AND DAGO PEAK STOCKS

Evidence from field studies suggests that assimilation, crystal fractionation, and multiple intrusion played important roles in the genesis of the Gem stocks. Assimilation is suggested by numerous inclusions of Belt rocks near their contacts with the Gem stocks. The inclusions range from those that did not react with the magma to ones that were almost entirely replaced by hornblende. The occurrence of Precambrian lead within the Gem stocks indicates that the Gem stocks interacted with the Belt rocks by assimilation or metasomatism. The presence of the mafic syenite with parallel potassium feldspar megacrysts and of the massive and monomineralic alkali-feldspar syenite, which have the appearance of cumulate and adcumulate phases respectively, supports crystal fractionation as an important process in the genesis of the intrusions. Finally, the cross-cutting relations suggest that multiple intrusion may have been involved.

ROAD LOG

Interval mileage is shown in parentheses. Mileages are to the closest 0.05 miles.

Spokane, Washington, to Wallace, Idaho

Mileage Description

0.0 Leave the intersection of Division Street and Interstate Highway 90 (I-90) in Spokane. Head east on I-90. *(30.7)*

30.7 Junction of U.S. Highway 95 (U.S. 95) and I-90. Continue east on I-90. Roadcuts east of this junction show Columbia River basalts on Precambrian Belt Supergroup rocks. Between Coeur d'Alene Lake and Wallace, Idaho, I-90 will remain in Precambrian metasedimentary rocks. *(36.4)*

67.1 Note the currently inactive smelter south of I-90 in Smelterville, Idaho. The devegetated area for the next few miles in the adjacent hills on either side of the road is the result of the fumes that were emitted from the smelter during the early years of operation.

From Smelterville to Wallace, I-90 passes through the Silver Valley and roughly parallels the Osburn fault zone. *(5.8)*

72.9 Note the statue of the miner north of I-90 at the Big Creek exit. This monument commemorates the death of 91 miners in the 1972 Sunshine Mine fire. *(7.5)*

80.4 Corner of 5th and Pine Streets in Wallace. Reset trip odometer to 0.0.

Wallace to Stop 7

0.0 At the corner of 5th and Pine Streets, turn north (left) onto Pine Street. Proceed north for one block to the T intersection of Pine and 6th Streets; the old Wallace train depot is straight ahead. Turn west (left) on 6th Street. Drive over the train tracks and head north on Dobson Pass Road, which parallels Ninemile Creek. Note the rocks of the Belt Supergroup that are exposed in the roadcuts on the way to the East Fork of Ninemile Creek. *(0.15)*

0.15 Look to the right and behind the old warehouse for a bit of local Wallace color, a yellow house which resembles a spaceship. Drive north on Dobson Pass Road. *(0.55)*

0.7 The Sierra Silver Mine tour area can be seen to the west (left). Many of the mines closed in this region in the early 1980s when silver and base-metal prices declined. Tourism has taken a greater emphasis in the local economy. One consequence is this family tour of an underground mine. *(2.3)*

3.0 The workings of the Dayrock Mine can be seen to the west (left). The main shaft of this mine intersects the Dobson Pass fault, which is one of the main north-south structures in this area. *(0.15)*

3.15 Turn east (right) at the Y intersection and head east along the East Fork of Ninemile Creek. *(0.25)*

3.4 Continue on paved road. The large roadcut ahead is in mafic syenite and is cut by the Dobson Pass fault at the west end. Note the old railroad grade at the top of the cliff. It will be the first stop. *(0.15)*

3.55 **Stop 1: Mafic syenite.**

Park at the turnout on the west (left) side of the road. Walk east up the ridge to the old railroad grade.

The mafic syenite is composed of a large variety of rock types and is intensely sheared. The contact along the Dobson Pass fault is marked by the occurrence of sheared, pink potassium feldspar that is heavily stained by iron oxide. This rock type can be seen at the west end of the large roadcut and along the railroad cut. Walk east along the old railroad bed. Notice the grain size and orientation of the potassium feldspars in the porphyritic mafic syenite. The aligned potassium feldspars strike from N. 60° W. to N. 85° W. and dip from 35° to 75° N. This strike nearly parallels the elongation of the mafic syenite unit (Figure 3). I believe that this suggests the orientation is a primary igneous feature and not the result of tectonic stresses. Interstitial quartz can be seen in mafic syenite samples.

Look also for the mafic clots composed of augite, hornblende, and plagioclase in the mafic syenite and for the gradational contact with the adjacent alkali-feldspar syenite. Aplitic hornblende syenite contains clots of equigranular mafic monzonite and zones of alkali-feldspar syenite. The mafic syenite is cut by dikelets of alkali-feldspar syenite up to 2 inches wide (5 centimeters). Walk approximately 0.2 mile along the abandoned railroad. Where the cut contains predominately alkali-feldspar syenite, turn around and return to the vehicle. Proceed south on the paved road to the gravel road that runs along the East Fork of Ninemile Creek. *(0.2)*

3.75 Go east (left) on the unpaved road that parallels the East Fork of Ninemile Creek. *(0.1)*

3.85 Note the cliffs of mafic syenite across the creek. The cliffs are below the same section that we viewed along the railroad. *(0.3)*

4.15 **Stop 2: Alkali-feldspar syenite.**

An excellent outcrop of alkali-feldspar syenite lies across the creek. This outcrop is heavily iron oxide stained probably because of the weathering of mafic minerals or iron sulfides. Continue northeast along the road. *(0.6)*

4.75 **Stop 3: Porphyritic monzonite.**

In the East Fork of Ninemile area, old mines commonly mark the contact of the Gem stocks with the Belt rocks. The dump ahead is at one of these contacts. The outcrop on the north side of the road is monzonite with potassium feldspar megacrysts. Note the epidote alteration. The porphyritic monzonite only occurs near the contact with Belt rocks and is probably the result of interactions between the stock and the country rocks. Continue northeast along the road. *(0.25)*

5.0 The road goes across one of the roof pendants present along the East Fork of Ninemile Creek. This contact zone between monzonite and Belt rocks is dominated by the development of hornfels in the Belt rocks. *(0.15)*

5.15 The trench north of the road provides an excellent exposure of the contact between banded green hornfels and the monzonite. *(0.05)*

5.2 The stock here is equigranular monzonite which lacks potassium feldspar megacrysts. *(0.2)*

5.4 **Stop 4: Equigranular monzonite.**

Turn south and cross the bridge. Park by the road on the south side of the creek. Walk northeast along the creek to an outcrop of unaltered, equigranular monzonite. Continue south on the road. *(0.1)*

5.5 Where the road forks, make a sharp right turn and go west to the top of the Success Mine dump. Note the change in float as we cross the contact between the monzonite and the Belt rocks. *(0.2)*

5.7 **Stop 5: Success Mine.**

This stop is at the top of the Success Mine dump. Features that can be seen include a contact between monzonite and Belt rocks and monzonitic samples with iron sulfides. In an outcrop at the southeast side of the dump, the contact of the stock is marked by recrystallization of the micas in the adjacent Belt rocks. Note the iron sulfide-bearing monzonitic rocks on the dump.

Theories about the origin of sulfides in the Gem stocks have always been controversial. Originally, the stocks were thought to have produced the mineralization, but it is now known that the Coeur d'Alene lead is predominately Precambrian (Long and others, 1960). Even Hershey (1917), who believed that the mineralization formed before the monzonite, conceded that at least some of the sulfides were deposited after the monzonite. After Precambrian ages were indicated for lead, some geologists continued to argue that the Gem stocks and a larger buried intrusion produced the ore in this area of the district by remobilization of Precambrian mineralization (Roales,

1973). Indeed, the Gem stocks contain disseminated sulfides and are cut by sulfide veins. Long and others (1960) stated that sulfides in the monzonite from the Success Mine probably resulted from remobilized Precambrian lead.

Several points can be made concerning the genesis of the Gem stocks. The presence of Belt rock inclusions in the stocks strongly suggests that Belt rocks were assimilated. If these metasediments contained sulfides, the sulfides would also have been assimilated. There is modern or Tertiary lead reported from the Sunrise and St. James prospects at the north end of the north Gem stock (Cannon and others, 1962). We may conclude that the lead in the Gem stocks is probably the result of some combination of both Precambrian and Cretaceous events. The fact that the lead is Precambrian does not necessarily mean the other metals are Precambrian. Given that the Gem stocks are Cretaceous, the age of the mineralization in the Gem stocks is probably younger than the age of the Precambrian lead. Retrace road back to the East Fork of Ninemile Road. *(0.25)*

5.95 Turn west (left) on the road that parallels the East Fork of Ninemile Creek. *(1.3)*

7.25 Take the south (left) fork at the Y intersection. The roadcut along here contains mafic syenite. For the next 2 miles, stay on the gravel road. The road will cross over the contact of the south Gem stock and the Belt rocks at several places. *(0.15)*

7.4 If the U. S. Forest Service gate is closed, use alternate Stops 6 and 7. If the gate is open, proceed ahead. *(0.1)*

7.5 This contact between Belt quartzite and the mafic syenite is probably a fault zone because the rocks are sheared on both sides of it and the contact lacks any metamorphic effects. *(0.6)*

8.1 The float indicates that the rock type here is monzonite. *(0.25)*

8.35 The rock type here is mafic syenite. *(0.25)*

8.6 This is a graditional contact between mafic syenite and monzonite. *(0.15)*

8.75 Belt rocks occur as float next to the road. *(0.05)*

8.8 Take the north (left) fork. This location is also the contact with the monzonite. *(0.2)*

9.0 The rock type here is mafic syenite. Since passing the U. S. Forest Service gate, the road has crossed the contact of the south Gem stock with the Belt rocks at several places, and the sequence has been the same each time. Belt rocks are separated from mafic syenite by monzonite. This suggests that the monzonite is a border zone of the intrusion. *(0.2)*

9.2 Take the north (left) fork. *(0.2)*

9.4 The rock type here is massive, alkali-feldspar syenite. There has been the same progression from mafic syenite to alkali-feldspar syenite as at Stop 1. *(0.1)*

9.5 The rock type grades from alkali-feldspar syenite to monzonite near the contact with a Belt roof pendant. *(0.1)*

9.6 The contact zone between monzonite and the Belt roof pendant is located here. *(0.1)*

9.7 **Stop 6: Sericitic alteration in the Gem stocks.**

The U.S. Forest Service gate is usually closed and locked. Park vehicle and walk northeast on the road. Take the north (left) fork at the Y junction and continue to the powerlines. Watch for the transition from a monzonite to sericitically altered rock of the same type. Sericite and iron oxides are found on fractures. Pervasive sericitic alteration is present in the vicinity of aplitic syenite dikes. Rare inclusions of Belt rocks are present. Return to the vehicle and drive back towards the East Fork of Ninemile Creek. *(0.3)*

10.0 **Stop 7: Mafic syenite.**

Sphene crystals and rare iron sulfides can be found here within the mafic syenite unit. Porphyritic syenite is present as float. Some of the mafic rocks are silicified. Continue towards the East Fork of Ninemile Creek. *(2.05)*

12.05 Turn south (left) on the road that runs parallel to the East Fork of Ninemile and return to Dobson Pass Road. *(0.25)*

12.3 Dobson Pass Road. Reset trip odometer to 0.0.

East Fork of Ninemile Creek to Dobson Pass

0.0 Turn north (right) on Dobson Pass Road. *(0.4)*

0.4 **Alternate Stop 7: Mafic syenite.**

Stop here if you cannot reach Stop 7. Park at the right side of the road and walk south along the road to the first roadcut. Observe the silicification in the mafic syenite and note the relation of the mafic syenite to the akali-feldspar syenite. Mafic syenite contains blebs, single crystals, stringers, and veinlets of alkali-feldspar syenite.

Return to the vehicle and continue north. The roadcuts are all in Belt rocks from here to the pass. The road crosses over the Dobson Pass fault at several places (Hobbs and other, 1965). *(0.8)*

1.2 Pull off the road to the right and stop. This is the road to alternate Stop 6. If you have visited Stop 6, reset trip odometer to 0.0. Go to the next 0.0 in the road log. If you have not visited Stop 6, go south (right) on this road. *(0.5)*

1.7 Turn right at the fork in the road. *(0.1)*

1.8 This is the contact zone of the Belt rocks with porphyritic monzonite. *(0.2)*

2.0 **Alternate Stop 6: Sericitically altered monzonite.**

Sericitic alteration is so intense in these rocks that the original rock type, which was monzonite with potassium feldspar megacrysts, is obscured. Quartz veins are also present. Continue north to the powerlines. *(0.05)*

2.05 Turn around at the switchback and head back to the Dobson Pass Road. *(2.05)*

4.1 Dobson Pass Road. Reset trip odometer to 0.0.

0.0 Turn north (right) on Dobson Pass Road. (All trip odometers should be at 0.0 whether you went to the alternate stops or not.) *(1.5)*

1.5 Dobson Pass. Again, reset trip odometer to 0.0.

Dobson Pass to Stop 13

0.0 The gravel road on the west side of Dobson Pass goes to Dago Peak. The Gem stocks exposed near Dago Peak probably represent the upper portion of the stocks that were transposed westward by the Dobson Pass fault (Hobbs and others, 1965). On the east side of Dobson Pass Road, the southernmost fork is a sharp right turn that will take you east toward Sunset Peak on a gravel road that is referred to below as the Sunset Peak road. The traverse is in Belt rocks. *(0.05)*

0.05 At this fork turn left toward Sunset Peak. *(0.45)*

0.5 Turn right at the fork and continue toward Sunset Peak. *(0.35)*

0.85 Pass the Rex Mine on the south (right) side of the road. *(0.25)*

1.1 Take the right fork and continue on the Sunset Peak Road. Note the contact between Belt rocks and monzonite. *(0.25)*

1.6 The road here passes through monzonite with mafic inclusions. *(0.25)*

1.85 Contact between monzonite and Belt rocks. *(0.05)*

1.9 The road cuts a short segment of monzonite. *(0.1)*

2.0 Pass road to right; this is difficult to see when overgrown with brush. Note its location however, because a later traverse will follow it. Continue on Sunset Peak Road. *(0.5)*

2.5 Pass fork to the left and continue straight to Sunset Peak. *(0.1)*

2.6 Pass another fork to the left and continue toward Sunset Peak. *(0.05)*

2.65 Pass fork to right. *(0.6)*

3.25 Pass road to right which is marked as Trail 137. Continue toward Sunset Peak. *(0.6)*

3.85 Turn right to Sunset Peak. The left fork goes to Goose Peak. *(0.2)*

4.05 **Stop 8: Overview of the Gem and Murray stocks.**

Park vehicle at the switchback and walk 0.1 mile to the top of Sunset Peak for an overview of the Gem and Murray stocks. At S. 35° W., the East Fork of Ninemile Creek can be seen, and at N. 40° E., Round Top can be seen. These two areas mark the southwestern and northeastern extents of the Gem and Murray stocks. Granite Gulch is at N. 65° E. and Tamarack Saddle at S. 60° E.: monzonitic rocks crop out at these locations. The alignment of these four points shows the strong northeast-southwest trend of the stocks.

The most mafic portion of the stocks (Stop 1) is at the southwestern end of the string of intrusions; the most felsic portion, the quartz, alkali-feldspar syenite, is at the northeastern end.

Note the small mine dump at S. 60° E. This is the Sunset prospect, which has galena in veins that cut the north Gem stock. The lead from these veins produces significantly younger model ages (Tertiary or Cretaceous) than the rest of the lead in the Coeur d'Alene district (Long and others, 1960; Cannon and others, 1962). Retrace road to Trail 137. *(0.8)*

4.85 Turn left onto Trail 137. *(0.2)*

5.05 Do not take the switchback to the right. Continue straight. *(0.15)*

5.2 Take the left fork and continue on Trail 137. *(0.1)*

5.3 This is the contact of porphyritc monzonite and Belt rocks. At the fork, follow Trail 137 to the left. This portion of the field trip requires a vehicle with above average clearance and should not be driven in passenger cars. *(0.2)*

5.5 Float in this area is predominately monzonite, but alkali-feldspar syenite is also present. *(0.2)*

5.7 **Stop 9: Alkali-Feldspar syenite with sulfides.**

Watch for a pile of rocks on the left side of the road. Park vehicle in the road and walk up the bank to the old prospect pit. Rocks from this pit are predominately alkali-feldspar syenite, but porphyritic and equigranular monzonites are also present. Quartz is present in the monzonite at this location. All of the rock types contain disseminated sulfides, and the potassium feldspars have a green tint that probably indicates the presence of sericite. Float of massive potassium feldspar rock, which is cut by monzonite, can be seen. *(0.1)*

5.8 **Stop 10: Equigranular monzonite.**

Turn around at the switchback and park the vehicle. Observe the float of monzonite with pink potassium feldspar crystals. Retrace route to the junction of Trail 137 and Sunset Peak Road. *(1.0)*

6.8 Go left on Sunset Peak Road and drive to the overgrown road at mileage 2.0 that was noted on the way to Sunset Peak. *(1.4)*

8.2 Turn left onto the dirt road. This road has a lot of brush, but is not as bad as it looks. This portion of the trip will compare and contrast a hornfels-type contact with one that shows evidence of assimilation of Belt rocks. *(1.0)*

9.2 **Stop 11: Contact zone of the Stock with Belt rocks.**

Park vehicle and walk along the road to observe the float that marks the contact between the Belt rocks and the Gem stocks. The float of Belt rocks shows recrystallization and coarsening of the micas and the development of banded and spotted hornfels. The float from the Gem stocks consists of sparsely porphyritic monzonite and rare hornblende-rich inclusions. Quartz veins and Belt inclusions are present in the monzonite float. Continue east on this road. *(0.2)*

9.4 **Stop 12: Inclusions in the monzonite.**

This monzonite outcrop is cut by many aplitic dikes. Drive past the outcrop and stop near its eastern margin. A large boulder that displays a large Belt argillite inclusion with bands of hornblende is on the south side of the road. The argillite inclusions probably reacted with the magma to form hornblende.

Walk northeast along this road and observe numerous inclusions that range from totally undigested argillite to those that are almost entirely replaced by hornblende and plagioclase. The abundance of mafic inclusions increases near the contact. Also note the abundance of potassium feldspar megacrysts, dendritic (plumose) feldspar veins, and alkali-feldspar syenite. Drive northeast on the road. *(0.3)*

9.7 Excellent huckleberry patches on either side of the road. *(0.2)*

9.9 **Stop 13: Brecciated contact of monzonite and Belt rocks.**

This stop is at the upper contact of the monzonite and Belt rocks. Note that the Belt rocks have been brecciated by monzonite and that the clasts have reaction rims of amphibole. Continue to drive northeast on this road. *(0.4)*

10.3 At this point porphyritic monzonite is present. *(0.1)*

10.4 Note the mine dump straight ahead. Take the switchback to the left and continue ahead to the Sunset Peak Road. *(0.1)*

10.5 Go straight ahead to the Sunset Peak Road. The switchback to the right leads to Tamarack Saddle along Trail 137. Note the contact of the Belt rocks and the monzonite at the switchback. *(0.2)*

10.7 Take the right fork of the Y intersection. *(0.2)*

10.9 Intersection of Trail 137 and the Sunset Peak Road. Turn left and retrace route to Dobson Pass along the Sunset Peak Road. *(3.25)*

14.15 Dobson Pass. Reset trip odometer to 0.0.

Dobson Pass to Spokane via Round Top

0.0 Turn north (right) at Dobson Pass and head toward Murray. *(0.8)*

0.8 **Stop 14: Lamprophyre dike.**

Note the dike with the spheroidal weathering in the roadcut of Belt rocks. Park at the turnout on the right side of the road at the switchback past the dike. This is one of the lamprophyres that are present throughout the district (Hobbs and others, 1965). Marvin and others (1984) indicate ages of 68.8±2.0 Ma to 50.8±1.5 Ma for lamprophyre dikes in this area. The dike is a very fine-grained lamprophyre and is about 2 feet (0.6 meter) wide. It contains numerous quartz inclusions. Continue north on the Dobson Pass Road. *(9.3)*

10.1 Turn right at the T junction and drive north on the Coeur d'Alene River Road. *(1.7)*

11.8 Turn east (right) on the Thompson Pass Road and go toward Murray, Idaho. *(3.6)*

15.4 Note the dredge piles on the south side of the road. These will be present along Prichard Creek until beyond Murray. Gold was discovered here before the rich silver and base metal veins of the Coeur d'Alene Mining District were found. Continue east on this road and through Murray. *(6.35)*

21.75 Note the entrance to Granite Gulch on the south side of the road. Granite Gulch contains a large mass of porphyritic monzonite. The roads into Granite Gulch have been been closed by the U.S. Forest Service. *(0.35)*

22.1 **Stop 15: Quartz, alkali-feldspar syenite.**

Park and walk along the road to observe float composed of quartz alkali-feldspar syenite. This phase is very consistent and almost everywhere cut by quartz veins. Return to vehicles and retrace route to the junction of the Dobson Pass Road and the Coeur d'Alene River Road. *(2.7)*

24.8 Entrance to the Golden Chest Mine on the north side of the road. This was one of the largest gold mines in the Murray district.

ACKNOWLEDGEMENTS

Field work for this project was supported in part by grants from the Washington Mining and Minerals Resource Research Institute and the Geological Society of America. I would like to thank Dr. Peter R. Hooper and Dr. Rolland R. Reid for their excellent and expedient reviewing of this field trip guide. Dr. Peter R. Hooper and Michael M. Bailey are acknowledged for pre-running the field trip.

REFERENCES

Armstrong, R. L., 1975, The geochronometry of Idaho: Isochron/West, No. 14, p. 1-50.

Cannon, R. S., Jr., A. P. Pierce, J. G. Antweiller, and K. L. Buck, 1962, Lead-isotope studies in the Northern Rockies, U.S.A., *in* A. E. J. Engel, H. L. James, and B. F. Leonard, editors., Petrologic Studies: A Volume in Honor of A. F. Buddington: Geological Society of America, p. 115-131.

Harrison, J. E., A. B. Griggs, and J. D. Wells, 1986, Geologic and structure maps of the Wallace 1° X 2° quadrangle, Montana, Idaho: Montana Bureau of Mines and Geology, Miscellaneous Investigation Series, Map I-1509-A, 2 sheets.

Hershey, D. H., 1917, Genesis of Success zinc-lead deposit: Economic Geology, v. 12, no. 6, p. 548-558.

Hobbs, S. W., A. B. Griggs, R. E. Wallace, and A. B. Campbell, 1965, Geology of the Coeur d'Alene district, Shoshone County, Idaho: U. S. Geological Survey Professional Paper 478, 139 pp.

Larsen, E. S., Jr., D. Gottfried, H. W. Jaffe, and C. L. Waring, 1958, Lead-alpha ages of the Mesozoic batholiths of western North America: U. S. Geological Survey Bulletin 1070-B, p. 35-62.

Lewis, R. S., T. H. Kiilsgaard, E. H. Bennett, and W. E. Hall, 1987, Lithologic and chemical characteristics of the central and southeastern part of the southern lobe of the Idaho batholith, *in* T. B. Vallier and H. C. Brooks, editors., Geology of the Blue Mountains Region of Oregon, Idaho, and Washington: The Idaho Batholith and Its Border Zone: U. S. Geological Survey Professional Paper 1436, p. 151-196.

Long, A., A. J. Silverman, and J. L. Kulp, 1960, Isotopic composition of lead and Precambrian mineralization of the Coeur d'Alene district, Idaho: Economic Geology: v. 55, p. 645-658.

Marvin, R. F., R. E. Zartman, J. D. Obradovich, and J. E. Harrison, 1984, Geochronometric and lead isotope data on samples from the Wallace 1° X 2° quadrangle, Montana and Idaho: U. S. Geological Survey Miscellaneous Field Studies Map MF-1354-G, 1 plate.

McDowell, R. W., 1971, K-Ar ages of igneous rocks from the western United States: Isochron/West, no. 2, p.1-16.

Parkinson, C. L., 1984, Geology of the Gem, Dago Peak, and Murray stocks, Coeur d'Alene Mining District, Idaho: University of Idaho M. S. thesis, 126 p.

Roales, P. A., 1973, Ore and monzonite dike relations, Monitor Mine, Coeur d'Alene district, Idaho: University of Idaho M. S. thesis, 144 p.

Streckeisen, A., 1975, To each plutonic rock its proper name: Earth-Science Review, v. 12, no. 1, p. 1-33.

The Geology and History of the Coeur d'Alene Mining District, Idaho

Earl H. Bennett [1]
Peter L. Siems [2]
James T. Constantopoulos[2]

INTRODUCTION

The Coeur d'Alene Mining District in northern Idaho (Figure 1) has the largest recorded silver production in the world. From the beginning of lode mining in 1884, the district's mines have produced over 1 billion ounces of silver, 8.5 million tons of lead, 3 million tons of zinc, and substantial quantities of antimony, cadmium, copper, and gold. The total value of this production is over $4.8 billion (Springer, 1987).

There are over ninety mines in the district. Eleven of these have produced over 3 million tons of ore. Of mining operations in the United States, the Coeur d'Alene contains the largest underground mine (the Bunker Hill, over 150 miles of workings), the deepest mine (the Star-Morning, over 7,900 feet deep) and the richest silver mine (the Sunshine, over 350 million ounces of silver). The district's mines annually produce about 16 million ounces of silver or about 40 percent of U.S. production.

Most of the mines are located along the South Fork of the Coeur d'Alene River and its major tributaries including Pine Creek, Ninemile Creek, and Canyon Creek. The South Fork is paralleled through the district by Interstate 90. Important towns in the district along I-90 from west to east include: Pinehurst, Kellogg, Osburn, Wallace, and Mullan.

GEOLOGY

The Coeur d'Alene Mining District mines are in metasedimentary rocks of the Belt Supergroup of Precambrian age. A description of the stratigraphic units in the Belt Supergroup (Figure 2) is given in Gott and Cathrall (1980) who used data from Hobbs and others (1965). The thickness of the Belt Supergroup in the dis-

[1] *Idaho Geological Survey, University of Idaho, Moscow, Idaho 83843*
[2] *University of Idaho, Moscow, Idaho 83843.*

Bennett, E. H., P. L. Siems, and J. T. Constantopoulos, 1989, The geology and history of the Coeur d'Alene Mining District, *in* V.E. Chamberlain, R. M. Breckenridge, and Bill Bonnichsen, editors, Guidebook to the Geology of Northern and Western Idaho and Surrounding Area: Idaho Geological Survey Bulletin 28, p. 137-156.

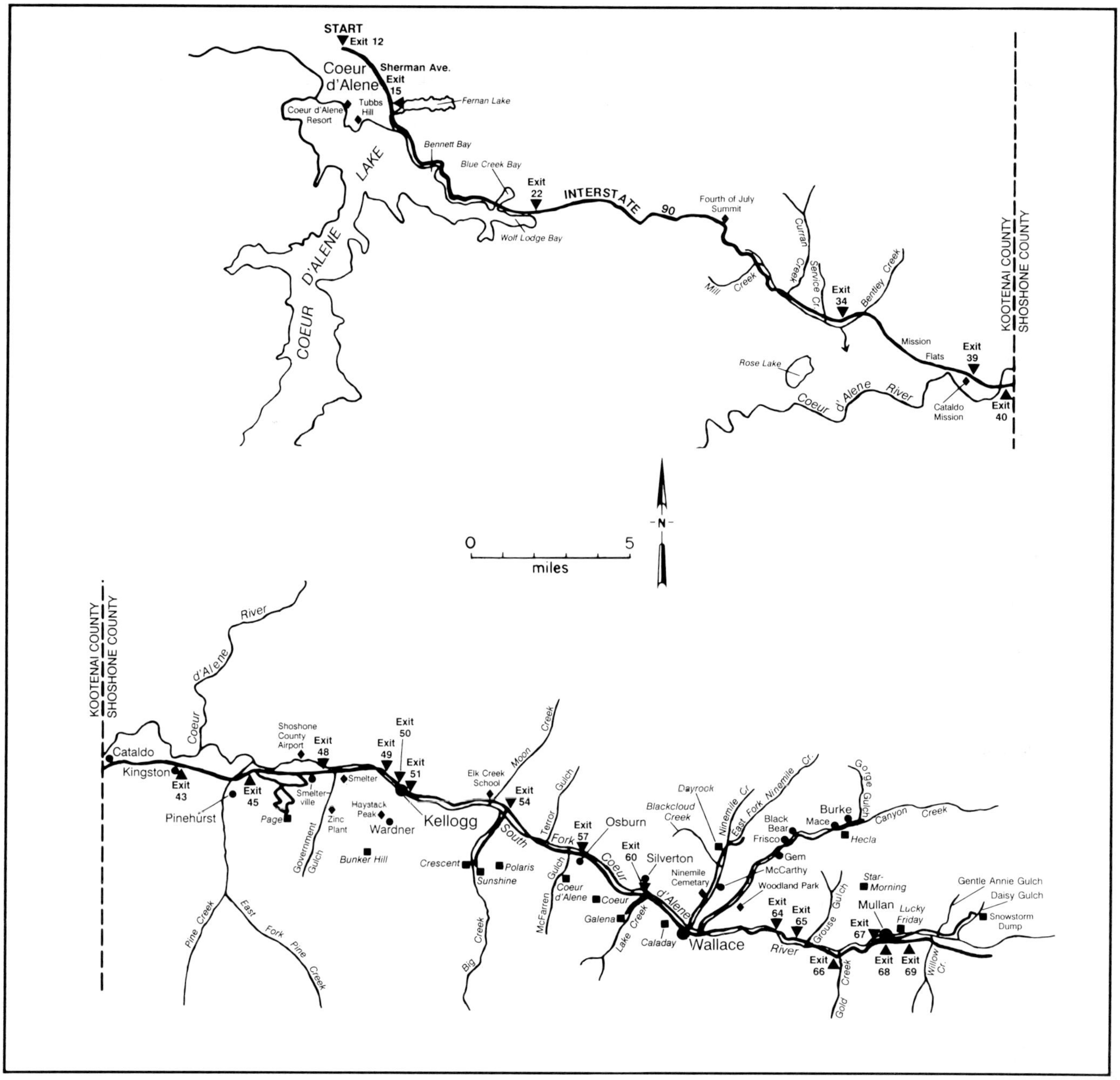

Figure 1. Road map of field trip-route along Interstate 90 from Coeur d'Alene to Mullan, Idaho, including the Coeur d'Alene Mining District.

trict is estimated to be at least 21,000 feet (Hobbs and others, 1965), although the base of the Prichard Formation is not exposed and part of the upper Belt has been removed by erosion. This thick sequence of Belt rocks is still only a relatively thin skin on a much thicker continental crystalline basement. The thin skin has been complexly deformed above a decollement (Harrison and others, 1980).

Intrusive rocks include the Gem stocks and Dago Peak stocks of Cretaceous age and diabase and lamprophyre dikes. Based on an aeromagnetic anomaly, an intrusive body, named the Atlas pluton, was inferred to exist below the Atlas Mine (Gott and Cathrall, 1980). However, the area is now known to contain enough magnetite and pyrrhotite to account for the anomaly.

The Coeur d'Alene district lies within the Lewis and Clark shear zone. The zone is composed of separate faults that collectively form a major intracontinental plate boundary in the northwestern United States (Reynolds and Kleinkopf, 1977). Major faults in the district that are

Group	Formation	Lithology	Thickness in feet
Missoula	Striped Peak	Interbedded quartzite and argillite with some arenaceous dolomitic beds; purplish gray and pink to greenish gray; ripple marks, mud cracks common, top eroded.	1,500+
	Wallace		4,500-6,000
	Upper part	Mostly medium- to greenish-gray finely laminated argillite, some arenaceous dolomite and impure quartzite, and minor gray dolomite and limestone in the middle part.	
	Lower part	Light gray, more or less dolomitic quartzite interbedded with greenish gray argillite, ripple marks and mud cracks abundant.	
Ravalli	St. Regis		1,400-2,000
	Upper part	Light greenish yellow to light green-gray argillite, thinly laminated; some carbonate-bearing beds.	
	Lower part	Gradational from interbedded argillite and impure quartzite at top to thick bedded pure quartzite at base; red-purple color characteristic, some green-gray argillite; some carbonate-bearing beds; ripple marks, mud cracks, and mud-chip breccia common.	
	Revett	Thin-bedded vitreous light yellowish gray to nearly white pure quartzite; grades into nearly pure and impure quarzite at top and bottom; cross stratification common.	1,200-3,400
	Burke	Light greenish gray impure quartzite; some pale-red and light yellowish gray pure to nearly pure quartzite; ripple marks, swash marks, and pseudo-conglomerate.	2,200-3,000
	Prichard		12,000+
	Upper part	Interbedded medium gray argillite and quartzose argillite and light gray impure to pure quartzite; some mud cracks and ripple marks.	
	Lower part	Thin-to thick-bedded, medium gray argillite and quartzose argillite, laminated in part; pyrite abundant; some discontinuous quartzite zones; base buried.	

Figure 2. Characteristics of the Belt Supergroup, Coeur d'Alene Mining District (adapted from Gott and Cathrall, 1980; stratigraphic data from Hobbs and others, 1965).

part of the Lewis and Clark zone include the Thompson Pass fault, the Placer Creek fault, and the Osburn fault. These structures all trend roughly east-west. Webster (1981) and Chevillon (1977) have suggested that soft sediment deformation in the Prichard Formation and other Belt sediments indicates that structures in the shear zone were active early in the history of the Belt basin.

The deformation of Belt metasediments within the Lewis and Clark zone is much more intense than elsewhere in the Belt basin, and folding and faulting are more intense in the Coeur d'Alene Mining District than elsewhere in northern Idaho. Thus, the most highly deformed parts of the Belt basin contain major mineralization. Folding and metamorphism in the district at about 850

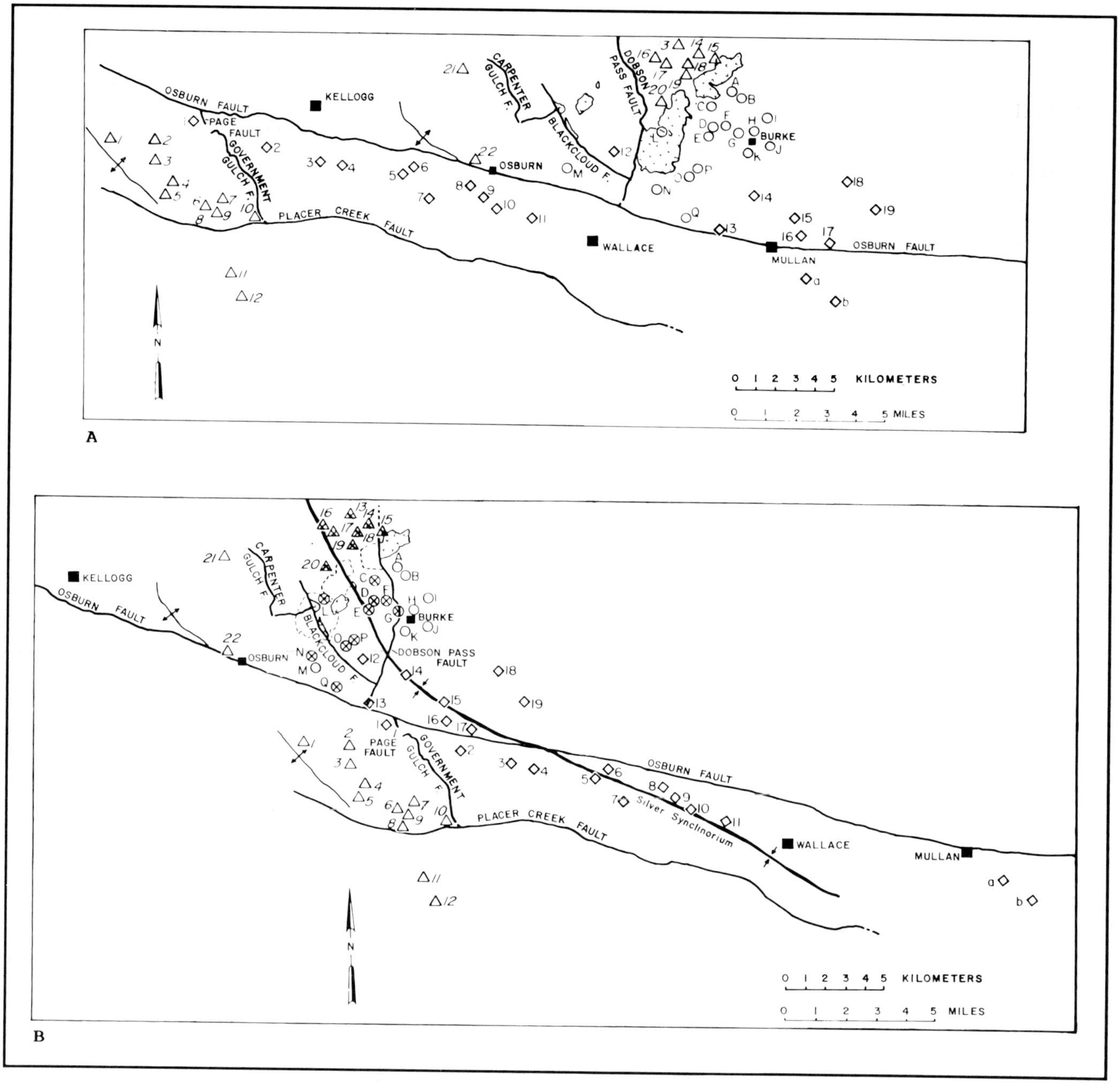

Figure 3. Location of ore deposits in the Coeur d'Alene Mining District before and after movement on the Osburn and Dobson Pass faults. (A) Present location of mines in the Coeur d'Alene district grouped by stratigraphic position. (B) Location of mines in the Coeur d'Alene district grouped by stratigraphic position as they would have been located prior to movement along the Dobson Pass and Osburn faults (Bennett and Venkatakrishnan, 1982).

m.y. ago have been attributed to the Kootenay orogeny (White, 1959; Harrison and others, 1974).

Effects of Faulting

The effects of major faulting in the district can be better understood by visualizing the district as divided into two blocks by the east-trending Osburn fault, with the northern half broken into two smaller blocks by the north-trending Dobson Pass fault. The location of ore deposits before and after movement on these faults is shown in Figure 3.

Hobbs and others (1965) conducted the most complete geologic study of the district, mapped most of the district, and described all of the major structures. They noted the displacement along the Dobson Pass fault and

Figure 3 continued.

MIDDLE PRICHARD DEPOSITS

Pine Creek Area
1. Hypotheek
2. Bobby Anderson
3. Amy Matchless
4. Sunset (Liberal King)
5. Lookout Mountain
6. Nabob
7. Sidney
8. Hilarity
9. Little Pittsburg
10. Highland Surprise
11. Douglas
12. Constitution

Upper Ninemile Creek and Sunset Peak Area
13. Idora, Tuscumbia, and Tough Nut
14. Silver Tip
15. Suset Lease
16. Carlisle
17. Amazon-Monitor
18. Nepsic
19. Interstate-Callahan
20. Rex (16 to 1)

Evolution District
21. Charles Dickens
22. Evolution

PRICHARD-BURKE DEPOSITS

A. Ambergris, Guelph, and Honolulu
B. Hercules
C. Tamarack
D. Union
E. Standard-Mammoth
F. Sherman
G. Hummingbird, Fairview, and Wide West
H. Tiger-Poorman
I. Ajax
J. Marsh
K. Hecla
L. Success (Granite)
M. Western Union
N. Canyon Silver (Formosa)
O. Helena-Frisco
P. Black Bear
Q. Golconda

ATLAS BLOCK

a. Atlas (Carbonate Hill)
b. Reindeer Queen

WALLACE DEPOSITS
15. Gold Hunter

REVETT-ST.REGIS DEPOSITS

1. Page
2. Bunker Hill
3. Alhambra

"Silver Belt"
4. Crescent
5. Sunshine
6. Polaris
7. Consolidated Silver (Silver Summit)
8. Coeur d'Alene (Mineral Point)
9. Coeur
10. Rainbow
11. Galena

Other Revett-St. Regis Deposits
12. Dayrock
13. Alice
14. Star-Morning
16. Lucky Friday
17. Vindicator

Stratabound Disseminated Copper-Silver Deposits
18. National
19. Snowstorm

Middle Prichard deposits are shown as △
Prichard-Burke deposits as ○
and Revett-St. Regis deposits as ◇
Mines below the Dobson Pass fault are shown by these symbols: ⊗ ▲
The Gem stocks are shown by a barbed pattern.

suggested that the Dago Peak stocks are the severed tops of the Gem stocks located east of the fault.

Hershey (1916) first recognized major strike-slip displacement along the Osburn fault by matching geologic units on both sides of the fault. He estimated that 10 to 15 miles of right-lateral displacement had occurred and noted that the area south of the Osburn fault near Kellogg must have originally been opposite a point just north of the fault near Wallace.

Hobbs and others (1965) indicated that about 26 kilometers of right-lateral displacement occurred along the Osburn fault from the Dobson Pass fault east to the state line and that about 12 miles of right-lateral slip occurred between the Dobson Pass fault and the town of Pinehurst to the west. The difference in offset is attributed to dip-slip along the Dobson Pass fault. Offsets of the Page-Government Gulch fault system from the Carpenter Gulch fault and of the Moon Creek anticline from the Pine Creek anticline were given as direct evidence of strike-slip movement along the Osburn fault.

Effects of Folding

Deformation in the district produced several large folds including the Big Creek anticline, the Deadman syncline, the Moon Creek and Pine Creek anticlines, and the Glidden Pass anticline. Several ideas have been proposed to explain the style of the deformation. Hobbs and others (1965) believed that all of the major folds were formed by a single event that produced north-south folds.

The north-south folds south of the Osburn fault were reoriented to a northwest trend by means of right-lateral movement. Juras (1982) presented evidence for two separate fold events, an earlier one forming north-south folds and a later one producing northwest-trending folds that were overprinted on the north-south folds. White (1989) reverses the order of these two events.

The line of mines south of the Osburn fault – beginning with the Crescent Mine and extending through the Sunshine, Silver Summit, Coeur, and Galena mines – is called the silver belt. Very rich silver-bearing ore is found in these mines. One of the large northwest-trending folds, the Big Creek anticline, is a major controlling structure for all of the mines, from the Page Mine through the Bunker Hill and silver belt mines. All of the silver belt mines are located on the overturned north limb of this anticline.

Description of the Ore Veins

Ore shoots in the mines are in steeply dipping veins that are classified as tabular replacement veins by Hobbs and Fryklund (1968). Most geologists believe that the veins are open-space fillings. The veins range from several centimeters to several meters in width. The veins are very persistent over long strike-lengths and also extend to great depths. The Star-Morning Mine was about 4,900 feet long and over 7,900 feet deep when mining ceased in 1982. The veins also show little vertical mineral zoning, although some veins like the Star-Morning do change along strike with the Star being zinc-rich and the Morning lead-silver rich. Postmineral deformation is evident in many of the veins.

The primary economic minerals in the district are galena, sphalerite, and tetrahedrite. Gangue minerals include quartz and siderite. Pyrite, chalcopyrite, and pyrrhotite are locally abundant. Fryklund (1964) and Hobbs and Fryklund (1968) discuss the paragenesis of the mineralization.

Most of the ore-bearing veins occupy one of three stratigraphic horizons (Figure 3). These horizons are (1) in the Prichard Formation near the middle quartzites (mines on Pine Creek and upper Ninemile Creek), (2) in the Prichard-Burke transition zone (mines surrounding Burke), and (3) in the Revett-St. Regis transition zone (Bunker Hill, Star-Morning, and mines in the silver belt). Sorenson (1968) credits Simos and Droste (1961) with first mentioning that the ore deposits in the Coeur d' Alene district were in three stratigraphic horizons. Bennett and Venkatakrishnan (1982) adopted this concept instead of using the familiar "mineral belts" described by Fryklund (1964). Individual mines in each group are indicated in the legend for Figure 3. The idea that all of the deposits are confined to the three stratigraphic horizons is a major key to understanding the genesis of the deposits. Bennett (1984) presented a hypothesis showing how all of the deposits in the district could have been generated from the three metal-bearing horizons.

Not all three groups of mines in the three horizons are equally important. Deposits are zinc-rich in the middle Prichard, lead/zinc-rich in the Burke-Prichard, and lead/silver/copper-rich in the most important Revett-St. Regis group.

The Bunker Hill and Star-Morning Mines

The Bunker Hill and Star-Morning mines are the largest ore producers in the district. The Bunker Hill Mine, unlike most of the other mines in the district, contains two distinct vein sets: a northwest-trending set of zinc/siderite veins, called the Bluebird veins, and a northeast set of silver/lead/quartz veins called the Link veins. The Star-Morning Mine can be visualized as a single vein (Figure 4). Several smaller veins also parallel the main vein. The Star-Morning is unusual in that it is lead/silver-rich in the Morning Mine (east end of the vein system) and zinc-rich in the Star Mine (west end of the vein system). Bennett (1984) suggested that the two vein systems in the Bunker Hill Mine could be explained by the mobilization of ore from the overlapping Revett-St. Regis and Burke-Prichard metal-bearing horizons. The zoning in the Star-Morning Mine could be the result of the mobilization of ore from adjacent Revett-St. Regis and Burke-Prichard horizons.

Age of the Deposits

Zartman and Stacey (1971) determined from lead isotopes that the age of some of the lead in the the Coeur d'Alene deposits was Precambrian and therefore unrelated to the Gem stocks (Cretaceous). Sericite has been dated by potassium-argon methods at 829 m.y. (Tallon orebody in the Bunker Hill Mine), 876 m.y. (the Lucky Friday Mine), and 447 m.y. (Galena Mine). The 800-850 m.y. dates may be close to the time of vein formation.

Origin of the Deposits

Some early researchers believed the district's veins were hydrothermal and related to the Gem stocks (Ransome and Calkins, 1908; Umpleby, 1917; Umpleby and Jones, 1923; Anderson, 1949; Fryklund, 1964; Hobbs and Fryklund, 1968; and Juras, 1982). The great Bunker Hill geologist, Oscar H. Hershey, proposed another idea: "I submit that the facts point to invisible diffused mineralization in the upper Prichard strata as the source of nearly all the lead and zinc in the deposits of the district" (Hershey, 1916).

Other supporters of a sediment-hosted stratiform or stratabound model for the origins of the metals included Sorenson (1968), Ramalingaswamy and Cheney (1982), Bennett and Venkatakrishnan (1982), Bennett (1984), Landis and Leach (1983), Leach and Hofstra (1983), and Leach and others (1988). Currently, a model that uses metamorphic fluids and derives metals from one or more stratabound/stratiform sedimentary horizons is most popular.

As a mechanism, metamorphism is appealing in that it solves the problem of the lack of mineral zoning in the deep veins. Bijak and Norman (1982) noted that temperatures based on fluid inclusion studies ranged from 320° C to 350° C in sphalerite and from 250° C to 320° C in quartz in samples from the Bunker Hill Mine. These temperatures are compatible not only with the assumed temperatures of the ore fluids but also with lower greenshist grade metamorphism, the grade of the host metasediments. Leach and Hofstra (1983) presented similar temperatures for other deposits in the district. Leach and others (1988) note that their data "clearly implicate regional metamorphic processes in the genesis of the ores. We suggest that the ore fluid was derived from metamorphic devolatilization processes and channelized along major faults and fractures." Leach and others (1988) further suggest that metals may have been derived from both the Prichard Formation and the Revett-St. Regis Formations as Bennett (1984) also suggested.

ROAD LOG

The road log travels east along Interstate 90 to the Coeur d'Alene Mining District located about 30 miles from the city of Coeur d'Alene, Idaho. The log is divided into four parts. The first part begins at the Sheraton Hotel in downtown Spokane, Washington, and ends at the city limits of Coeur d'Alene, Idaho. The second part takes up at Coeur d'Alene and goes to Pinehurst. The third part, which covers the Coeur d'Alene Mining District, begins at Pinehurst and ends at Mullan. The fourth part consists of Side Trips into the mining district that depart from the main I-90 log. Except for the Side Trips, mileage is accumulated from the Sheraton Hotel in Spokane.

Part 1: Spokane, Washington, to Coeur d'Alene, Idaho

This part of the road log begins at the Sheraton Hotel in Spokane. Set tripmeter at 0.0. From the entrance to the Sheraton turn right on Spokane Falls Boulevard. Immediately move over to left lane. Turn left on Bernard Street. After one block, turn right on Browne Street. Move over to left lane. Take the entrance on your left to I-90 east and to Coeur d'Alene.

Mileage	Description
1.8	Ridge and cliffs on the right of I-90 are outcrops of Columbia River basalt.
5.2	The summit of Mount Spokane is on the skyline ahead on the left. The summit and entire western flank of the mountain are underlain by granitic rocks of the Spirit Lake pluton of the Colville batholith.
8.0	Hills ahead on the right are underlain by Prichard Formation(?) metasedimentary rocks of the Belt Supergroup.
9.2	Cliffs of Precambrian gneiss (pre-Belt or Prichard?) on left.
10.0	Kaiser Aluminum rolling mill on left. The Pacific Northwest region is an important producer of aluminum because of the availability of hydroelectric power from the Columbia River and its tributaries.
19.4	Sign: Welcome to Idaho
23.5	Post Falls, Idaho, is on the right. The town is named for Frederick Post who purchased the site from the Coeur d'Alene Indians in 1871. Post dammed the river and built a sawmill. He sold the site to R.K. Neill in 1900. Neill wanted to build a new dam and power station to provide electricity to the Coeur d'Alene district mines, but he sold out to Washington Water Power Company (WWP). WWP built a new dam and completed a powerline to the Coeur d'Alene district in 1903. The district's power was orginally generated at Spokane Falls until the station at Post Falls was completed in 1906 (Conley, 1982). Mountains on left, across the Rathdrum Prairie, are underlain by Precambrian high-grade gneiss and cupolas of the Spirit Lake pluton.
25.1	Flat hill ahead is underlain by glacial flood deposits.
27.2	Rest area on I-90.
29.2	Entering city of Coeur d'Alene.

Part 2: Coeur d'Alene to Pinehurst, Idaho

Mileage Description

31.2 Exit 12. Junction with U.S. Highway 95. Stay on I-90. The Interstate follows the old roadway of the Mullan road named for Captain John Mullan. Mullan was charged with building a military road from Fort Walla Walla in the Washington Territory to Fort Benton in present-day Montana. The road was completed to Cataldo in 1862, but frequent flooding by the Coeur d'Alene River made the route almost impassable.

32.6 Columbia River basalt in roadcut on left.

33.5 Milepost 15. Exit 15, Sherman Ave. Fernan Lake on left.

In 1878, General William Tecumseh Sherman selected part of the present-day city of Coeur d'Alene for a military fort, whereupon the U.S. government set aside 999 acres for the compound to be called Fort Coeur d'Alene. Part of the original site is now occupied by North Idaho College. In 1887, the installation was renamed Fort Sherman in honor of its founder.

The name Coeur d'Alene had an earlier origin. When David Thompson, an early-day explorer, visited the area in 1809, the Indians were known as the Coeur d'Alene or "Awl Heart," a name that cynically described the sharp trading practices of the Indians who, in the eyes of their critics, had hearts as small as the point of an awl (Conley, 1982).

The discovery of the mines in the Coeur d'Alene district, 30 miles to the east, spurred the economic development of the town that sprang up near the fort. The town became the supply center for the mines and was incorporated as Coeur d'Alene in 1887. Fort Sherman was abandoned as a military post in 1900 (Conley, 1982).

34.4 South of I-90 is the former site of the Potlatch Company lumber mill that was recently torn down. A new golf course for the Coeur d'Alene resort may be built here. West of the mill site is Tubbs Hill, named for Tony Tubbs who subdivided his homestead to make part of the new city. The west side of the hill was the docking point for paddle-wheel steamers that transported men and supplies across the lake to the landing near the Cataldo Mission. Ore being shipped from the district was unloaded at a railhead near Tubbs Hill for transport to various smelters by the Northern Pacific Railroad.

36.0 Turnoff, "Historical Site 268." The sign tells of the lake steamers that transported ore and people on Coeur d'Alene Lake. The most famous of the steamers was the *Georgie Oakes* (named for the daughter of the president of the Northern Pacific Railroad). Built in 1890, the beautiful sternwheeler plied the lake's waters until 1927 when, as part of a Fourth of July celebration, it was set afire, a common fate of worn-out steamers.

36.1 Milepost 17.

36.4 Bennett Bay. Coeur d'Alene Lake, one of the most beautiful lakes in Idaho, formed when flood gravels from glacial Lake Missoula dammed the St. Joe River. The past pollution of the lake by mill tailings from the old mining operations in the Coeur d'Alene district and the present-day economic development of the lake are major environmental concerns for the people of northern Idaho.

41.2 Exit 22. Harrison-St. Maries Exit. Wolf Lodge Bay District.

45.9 Roadcuts in the Prichard Formation.

47.8 Fourth of July Pass (elevation 3,070 feet). Prichard Formation along road. Overpass crosses I-90. Just east of the crest is the site where the Mullan tree stood until recently; the tree was a survey marker on the old Mullan road. Fourth of July Canyon was named by Mullan's men when they celebrated the holiday near here in 1861.

51.4 Milepost 32. A little west of the State Department of Transportation weigh station, and just west of Curran Creek. Rocks are Prichard Formation. A fault at Service Creek, south of the I-90, separates lower Prichard rocks to the west from upper Prichard units to the east. The Osburn fault is about 1 ½ miles north of I-90.

54.5 Milepost 35. The hill south of Bentley Creek is composed of the Prichard Formation (west part) and the Burke Formation (east part). Past here are the bench gravels and alluvium of the flood plain of the Coeur d'Alene River and the old mill tailings of Mission Flats. Mining companies, worried about pollution, operated a dredge at Mission Flats from 1930 to 1960 to remove old mill tailings from the Coeur d'Alene River.

58.6 Milepost 39. Old Mission State Park turnoff. Hills to north are Prichard Formation.

The Cataldo Mission of the Sacred Heart is the oldest standing building in Idaho and a registered national historical landmark. The mission was built in 1853 by the Coeur d'Alene Indians under the direction of a Jesuit missionary, Father Anthony Ravalli, and named for another priest, Father Joseph Cataldo. This structure replaced an earlier mission near St. Maries.

59.7 Milepost 40. Exit 40, Cataldo turnoff. The town of Cataldo is named for the Jesuit priest Joseph Cataldo, who arrived among the Coeur d'Alene Indians in 1865. He was still active with the tribe when he was 90 years old.

60.0 Kootenai-Shoshone County line. Hill just east of Cataldo is underlain by the Burke Formation.

Shoshone County and the local Indian tribe were mistakenly named by the Washington territorial government. Knowing little of the geography of this wild place, the legislators thought the local Indians were part of the Shoshone-Bannock tribe of southeast Idaho. Therefore, the name Shoshone County was given to the new mining area.

62.4 Exit 43, Kingston. Hill to the left (north) is Kingston Ridge (Burke Formation). Hill to the south is underlain by Prichard Formation. The two hills are separated by the trace of the Osburn fault, which is in the valley. This location is the confluence of the North and South Forks of the Coeur d'Alene River.

It is likely that one of Captain Mullan's men discovered gold in the South Fork of the Coeur d'Alene River. The initial discovery, however, is credited to Andrew Prichard who found gold at Evolution (near Osburn) in 1878. Prichard's later discovery on the North Fork of the Coeur d'Alene River near Murray in 1881 was the spark that ignited a gold rush to the area. The rush was further fueled by the Northern Pacific Railroad, which had recently completed its line to Spokane Falls. In 1884, the company plastered the entire country with handbills touting the riches of the Murray gold fields to drum-up business for the new line. Thousands flocked to the area but the boom died a short while later.

The initial gold rush to Murray was over by 1885, but by then prospectors had fanned out in all directions. On May 22, 1884, John Carten and Almeda Seymour staked a showing of galena at the site of the Tiger Mine. This discovery is credited with being the first in the district.

In 1917, a large bucket-wheel dredge was assembled to work gold-bearing gravels on the North Fork. The dredge was used through the 1920s. Piles of gravel or dredge spoils are heaped near the town of Murray.

Part 3: Pinehurst to Mullan, Idaho: the Coeur d'Alene Mining District

Mileage Description

64.7 Milepost 45. Exit 45, Pinehurst and the old road to Smelterville. The Bunker Hill smelter stacks can be seen from I-90, just east of the exit. Pine Creek is in the Prichard Formation. The Osburn fault passes from the north side to the south side of I-90 at Pinehurst. The town is the gateway to the Silver Valley and the Coeur d'Alene Mining District.

Located in the Pine Creek drainage are several mines that produced zinc-rich ore. These are vein deposits in the Prichard Formation above the white quartzites. The largest producing mine was the Sidney that was developed in the late 1920s to provide zinc feed for the Sullivan Mining Company's (a joint venture between Bunker Hill and Hecla) new electrolytic zinc plant. A tramway carried the ore from the Sidney to the old Sweeney mill located just west of the present-day smelter. Other mines on Pine Creek included the Constitution, Highland-Surprise, Little Pittsburgh, Liberal King, Douglas, and Nabob. Beautiful samples of stratabound ore, reminiscent of samples from the Sullivan Mine in Kimberly, British Columbia, can be found on the Constitution Mine dump. A reconstruction of movement along the Osburn fault brings the Pine Creek mines into a north-south alignment with the mines at the head of Ninemile Creek. Both sets of mines are in the same stratigraphic horizon and were also primarily zinc-producing operations.

For Side Trip 1: Pinehurst to Page and the Page Mine – see Part 4: Side Trips.

65.9 South of I-90 is the Page Mine's tailings pond. The Page Mine is a little further south. The Page was operated by the Federal Mining and Smelting Company (later Asarco) from 1924 to 1969. The surface plant at the mine burned in 1973.

67.1 Exit 48, Smelterville. Beginning of smelter slag pile (black material). Rocks are the Prichard Formation south of the valley and the Burke Formation north of the valley. The drainage to the south is Government Gulch, the location of Bunker Hill's zinc plant, fertilizer plant, and laboratories.

67.8 The Bunker Hill Company's smelter (Figure 4) is south of I-90. The rocks to the north are Prichard Formation all the way to the east side of Osburn.

69.2 Exit 49, Bunker Avenue turnoff. Bunker Hill tailings impoundment to the right.

69.8 Exit 50, Hill Street/Silverhorn exit. The Bunker Hill concentrator is south of I-90. Behind the concentrator is the Kellogg tunnel, entrance to the Bunker Hill Mine. The Silverhorn ski area is on top of the ridge above Kellogg. Construction will soon start on a gondola that will connect Kellogg with the Silverhorn ski area. When completed, this will be the longest gondola in North America.

The site of the Bunker Hill Mine was dis-

Figure 4. Bunker Hill smelter.

covered by Noah Kellogg in 1885. A famous fable claims that the mine was really found by Kellogg's jackass. The first major development of the mine was by Simeon Reed, a Portland, Oregon, businessman, who purchased the mine in 1887 and formed the Bunker Hill Mining and Concentrating Company. The mine was originally connected to the old South mill by a tramway that passed over Wardner and Haystack Peak. In 1903, the two-mile long Kellogg tunnel was completed. The discovery of the huge March orebody in 1904 assured success for the venture. The old South mill had been blown up in the mining labor war of 1899. Several mills have operated since, the latest located east of the smelter. The company built its own smelter in 1917 and the electrolytic zinc plant (with Hecla) in Government Gulch in 1928. Both plants underwent substantial changes over the years, including the addition of the 715-foot-high stack at the smelter and the 610-foot-high stack at the zinc plant completed in 1978. The company was instrumental in developing uses for zinc and contributed much to the zinc diecasting industry. In later years, sulfuric acid generated from the smelter was combined with phosphate ore from southeastern Idaho to make fertilizer in another plant in Government Gulch. The mining and metallurgical complex was the lifeblood of the valley until Gulf Resources and Chemical Corporation (who had purchased it in 1968) closed the operation in 1981 with the loss of 2,100 jobs. The closure was a devastating economic blow to the Silver Valley, and recovery is still underway. When the complex closed, it was producing about 20 percent of the nation's refined lead and zinc and 25 percent of its silver. The smelter and zinc plant will probably never operate again. The mine, however, was reopened in 1988 and once again ore flows from the Bunker Hill. The mine contains over 150 miles of underground workings and is the largest lead/zinc mine in the United States. An impressive model of the mine can be viewed at the company's engineering office by prior arrangement. The old company staff house is now home for the Kellogg Mining Museum, also a worthwhile stop.

The geology of the Bunker Hill Mine is complex. Unlike other mines in the Revett-St Regis Group, mineralization at Bunker Hill was exposed at the surface. Two major vein sets are identified in the mine. One is called the Bluebird veins, a northwest-trending, zinc-rich set with siderite gangue; the other is called the Link veins, a northeast-trending, silver/lead-rich set with quartz gangue. The Lucky Friday Mine may be in one of the Link veins that has been displaced to the east along the Osburn fault. A series of reverse faults, including the well-known Cate fault, are also important in the mine. Near the glory hole above Wardner is the Tyler Ridge flexure that may be either the expression of the Big Creek anticline in this area or a subsidiary fold on the anticline. The latest mining effort is located in the Quill orebody, a zinc-rich part of the mine.

70.1 Exit 51, Division Street/Wardner exit. Haystack Peak can be identified by the sign on its top. The town of Wardner is named for Jim Wardner, one of the earliest promoters in the district and a principal owner in the original development of the Bunker Hill Mine. Above Wardner is the glory hole, the original discovery site and old workings of the mine, and the old dumps of the Last Chance Mine.

71.6 Milepost 52. Elizabeth Park Road. Large outcrop along the road. Prichard Formation north and south of I-90.

73.2 Moon Gulch, Elk Creek School. The Prichard Formation is north and south of I-90. Moon Gulch is near the axis of the Moon Creek anticline, which is part of the Pine Creek anticline that is offset by the Osburn fault.

73.3 Milepost 54. Exit 54, Big Creek Exit. The Polaris mill is visible ahead, on the west end of Osburn.

Left of the turnoff from I-90 is an iron sculpture of a raise miner by the well-known mining artist Ken Lonn. The statue commemorates the 91 miners who died in the tragic Sunshine Mine fire of 1972. This was the worst fire in a hard rock mine in U.S. history. Most of the fatalities were due to carbon monoxide poisoning. As a consequence of the fire, new and improved safety standards were instituted for underground mines.

For Side Trip 2: Up Big Creek to the Sunshine Mine (Figure 5) — see Part 4: Side Trips.

74.5 North of I-90 is the site of the Evolution Mine near where Andrew Prichard discovered gold in 1880.

75.1 Silver Dollar Mine dump on right. This is close to a fault contact between the Prichard Formation to the west and Wallace Formation to the east. The Wallace Formation is south and the Prichard Formation north of I-90.

75.6 South of I-90 is Rosebud Gulch. The Polaris mill is on the east side of the gulch. The mill is near the site of a $17 million program by the Consolidated Silver Corporation to look for deep orebodies in the old Silver Summit Mine. Some ore was discovered but not in quantities economic to mine or process. Terror Gulch is north of the highway. North of I-90 is all Prichard Formation rocks.

Figure 5. Sunshine Mine.

75.8 Milepost 56. Outskirts of Osburn. Established in 1887, the town is named for Billy Osburn, a principal owner in the early Mineral Point (later Coeur d'Alene) Mine. To the right of I-90 is McFarren Gulch and the mill of the old Coeur d'Alene Mine (not the same mine as the Coeur Mine which is on the east end of Osburn). The Coeur d'Alene Mine was a major producer of copper and silver in 1941-1942, but after that the ore was quickly exhausted. Coeur d'Alene Mines Corporation was formed in 1928 and is a major mining company, today owning the Coeur Mine and the Rochester Mine in Nevada (the largest heap-leach silver mine in the world) among other properties.

76.6 Exit 57, Osburn. Prichard Formation north of highway, Wallace Formation south.

77.2 South of the road is the Zanetti Brothers Cement plant.

In the past, all of the mills in the district dumped mill tailings directly into the Coeur d'Alene River and its tributaries. Periodic flooding moved the metal-rich tailings through the drainage and eventually into Coeur d'Alene Lake. This practice contaminated the Silver Valley and the lake with metals. Large wooden tailings dams were built at Woodlawn (on Canyon Creek) and near Osburn and Pinehurst. These dams were frequently breached by spring runoff. In 1944, Hecla Mining Company built a sink-and-float plant near Osburn to reclaim zinc and lead from about 2 million tons of old mill tailings that filled the valley. The plant produced substantial amounts of the metals but was destroyed by fire in 1948.

Ahead on the left is the tailings pond for the Coeur Mine. Today, all of the active mines impound tailings, and no contaminated water is discharged into the river system. The Osburn fault passes across the valley and is now north of I-90. The fault trace goes through the prominent saddle to the northeast.

77.9 The Coeur Mine (Figure 6) is south of the Silver Hills Junior High School and an old overpass. Rocks on the lower slopes are part of the Wallace Formation. Units of the St. Regis Formation form the top of the hill. The units are separated

Figure 6. Coeur Mine.

Figure 7. Galena Mine.

by the Polaris fault. The hill north of I-90 is underlain by Wallace Formation, which also outcrops on both sides of the road leading into the city of Wallace.

The green mill buildings and headframe of the Coeur Mine are visible from I-90. The mine's tailings impoundment is to the north. The Coeur was placed in production in 1974 and is the newest mine in the Silver Valley. It is owned by Coeur d'Alene Mines Corporation and operated by Asarco on a long-term lease. The Coeur produces about 2.5 million ounces of silver a year.

79.2 Exit 60, Silverton, Shoshone County Hospital, U.S. Forest Service headquarters.

For Side Trip 3: Up Lake Creek to the Galena Mine—see Part 4: Side Trips.

79.6 Long Creek. The turnoff to the Galena Mine (Figure 7) is about 1 mile up Long Creek from the old Wallace-Osburn road. At the mine the St. Regis Formation is faulted against the Wallace Formation along the Polaris, Argentine, and Kilbuck faults.

The Galena Mine was purchased by a subsidiary of the Callahan Lead-Zinc Company (today's Callahan Mining Corporation). The property was mined until the Great Depression. Asarco decided, after a geologic study of the area, that the mine had potential for deep development, and in 1947, the company arranged a lease agreement with Callahan. A shaft sunk almost 3,000 feet found commercial silver ore. This new Galena Mine was placed in production in 1955 and has had the largest cumulative production over the past 15 years of any mine in the district. This success is due to Asarco's efficient operation and to over 22 years of labor harmony at the mine.

80.6 The concrete footings of the old Hercules mill are on the left. The Hercules mill was built in 1911 and burned in 1976. Ore from the Hercules Mine was processed here as well as from other Day family holdings (see Side Trip 5). All rocks in this part of the valley are part of the Wallace Formation.

80.8 Milepost 61. Entering Wallace. Last outcrop south of I-90 is Wallace Formation. NOTE: new construction on I-90 will bypass Wallace. After construction is completed, visitors should take the Wallace exit to continue on this road log.

81.5 Bank Street, Coeur d'Alene District Mining Museum, Asarco office, Whiter and Bender Building, Rossi Insurance Building, and County Courthouse.

81.6 Intersection of Bank Street and 7th Street.

For Side Trip 4: Wallace Up Ninemile Creek—see Part 4: Side Trips.

For Side Trip 5: Wallace to Burke—see Part 4: Side Trips.

81.7 Proceed past the Public Safety Building on Bank Street (also the present route of I-90).

81.8 Bank Street, State Highway 4, (Canyon Creek road) intersection. The Wallace Formation is exposed in new roadcuts for I-90 just to the east of Canyon Creek.

84.0 Golconda Mine site to the left. The mine closed in 1957. The Wallace Formation is north of the highway; the St. Regis Formation is to the south.

84.3 Exit 64, Golconda District.

84.8 Rock Creek property, owned by Hecla Mining Company on the right side of the road. A 5,000-foot-deep drill hole, the deepest ever drilled in the Coeur d'Alene Mining District, was completed at Rock Creek in 1988 by Teck Inc.

85.4 Exit 65, Compressor District, Grouse Creek. St. Regis Formation on both sides of I-90.

The Compressor District was not really a mining district at all but the site of the largest water-driven air compressor in the world. Built in 1900 and used until 1950, the compressor was driven by a 33-foot-diameter Pelton wheel flanked by two 11-foot-diameter wheels. The machine produced over 1,000 horsepower and provided all of the compressed air for the large Morning Mine. Streams were diverted to the site from many miles around to drive the compressor. Eventually, electric motors were added to help run the unit in the dry part of the year. Commercial electric power from Spokane arrived in the Coeur d'Alene Mining District from Spokane in 1903. At the time this powerline was one of the longest in the world.

86.6 Exit 66, Gold Creek exit. St. Regis Formation.

86.9 Local sign noting "Arnolds fountain." Gold Creek Mining Company. The Morning dumps are to the north around the next corner. The water in the fountain is fed by part of the system that flumed surrounding streams to the Grouse Creek compressor.

87.4 Exit 67, Morning District. The yellow buildings on the left are all that is left of the Morning mill. The portal of the Morning No. 6 tunnel is in rocks of the Wallace Formation. The 2-mile-long tunnel crosses the Osburn fault.

The Morning Mine (Figure 8) was developed in 1889 by Charles Hussey, a Spokane banker. After a change in ownership in 1897, a new mill was built and the No. 6 tunnel was driven from the mill site near Mullan to the mine. In 1905, the mine was purchased by the Federal Mining and Smelting Company, which merged with Asarco in 1953. The Morning mill was an active laboratory where Federal's metallurgists wrestled with the problem of separating zinc and lead from the complex ore. Selective flotation, discovered in the 1920s, solved this problem. In 1961, Hecla obtained a lease on the mine and in 1966 purchased the property. The Morning mine and Star mine (both part of the same orebody, see Side Trip 5) were then operated as the Star-Morning unit. Fire destroyed the Morning mill in 1957. Hecla ceased mining the Star-Morning in 1982.

The Star-Morning Mine is essentially a one-vein system with a few parallel veins. The mine is about 4,900 feet long and 7,900 feet deep. The deposit is in the Revett-St. Regis group of mines but is different than the other mines. The Morning Mine on the east end of the system is lead/silver-rich whereas the Star Mine is zinc-rich. The Star-Morning has the second largest production in the district after the Bunker Hill. Both the Star-Morning and the Bunker Hill mines are anomalous in comparison with other Revett-St. Regis group mines that contain a simpler mineralogy and lower total production.

88.0 Exit 68, Mullan Exit, large letter "M" on hillside above Mullan. At Mullan, I-90 crosses the Osburn fault.

88.4 Third St. Ramp over I-90. Ahead is the Silver shaft of the Lucky Friday mine. Rocks are in the St. Regis Formation.

Located to the north of I-90 is the football field of Mullan High School. This field is near the site of the Gold Hunter mill. The Gold Hunter was developed by Day Mines, Inc. The mine is in the Wallace Formation as were the early workings of the Sunshine Mine. As is typical of these deposits, the ore was rich but spotty. In 1977, Hecla decided to look below the Gold Hunter mine, to probe for the Revett-St Regis horizon at depth. The company drove a drift under the old workings from the 4050 level of the Lucky Friday. At that depth the rocks are still part of the Wallace Formation. This indicates that there is a thick sequence of Belt rocks folded in the Deadman syncline between the Big Creek anticline to the south and the Glidden Pass anticline to the north. Bennett and Venkatakrishnan (1982) called this area the Silver synclinorium and estimated that at least 10 miles of horizontal Belt rocks are folded into a zone now about 2 miles wide.

88.7 Exit 69, East Mullan/Lucky Friday Mine. From the exit, turn left and go across the overpass to the Lucky Friday Mine (Figure 9). The Lucky Friday tailings impoundment is to the left of the overpass. The Atlas Mine is to the right of the overpass and is in the St. Regis Formation.

Like the Sunshine and Galena mines, the Lucky Friday was a late bloomer in the district.

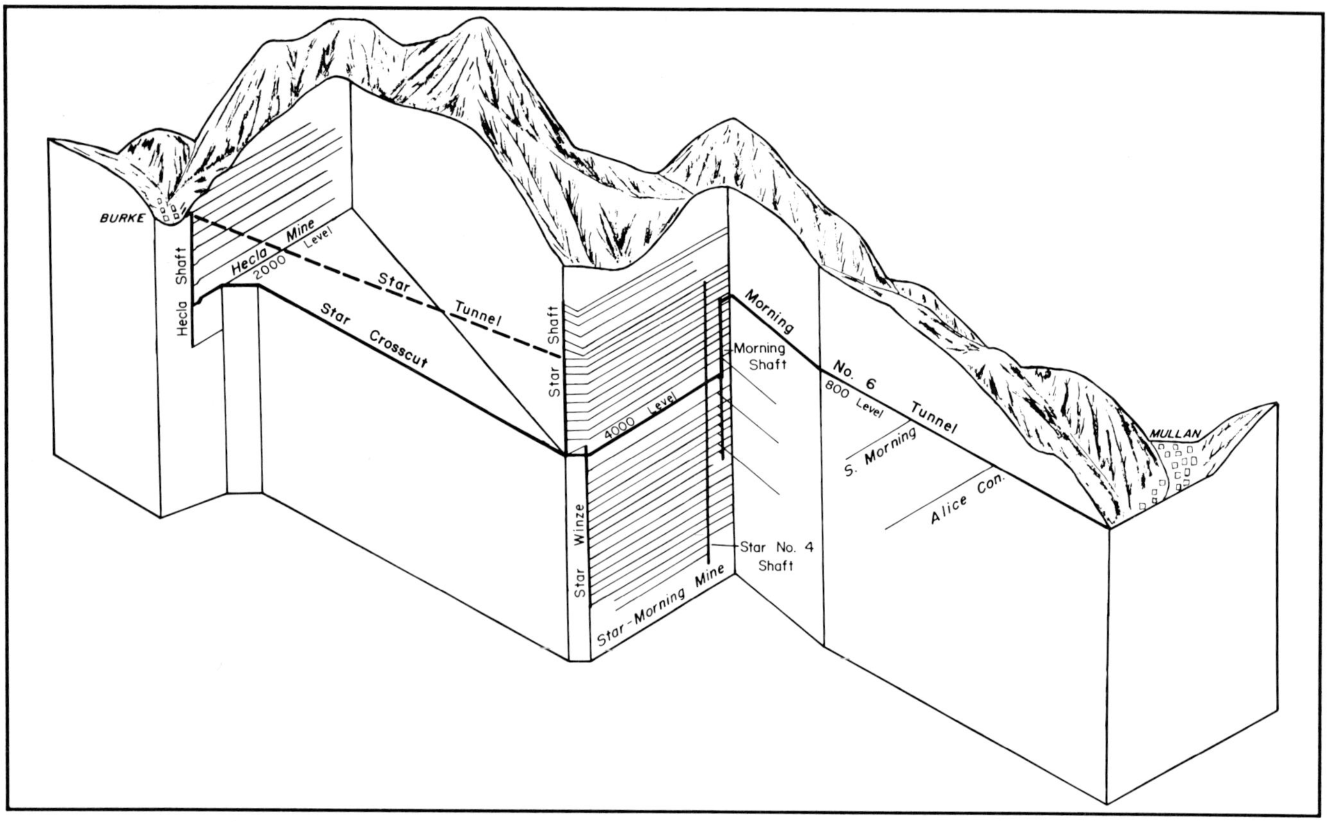

Figure 8. Projection of underground workings between Burke and Mullan.

Miners had to sink through upper Belt rocks to get to the favorable Revett-St. Regis horizon. In 1912, the Friday was sold at a sheriff's sale for $2,000 and again in 1936 for $120 in back taxes. In 1938, John Sekulic, a garage mechanic in Mullan, purchased the mine for $15,000 and formed the Lucky Friday Mining Company. With a lot of faith in the venture and enough technical and financial help from Judge Albert Featherstone, the president of Golconda Mining Company, mining began in earnest. The company shipped its first ore in 1942. Like the Sunshine, the ore got better and better as the shaft went deeper. Hecla Mining Company started buying stock in the Lucky Friday, and in 1964 the two companies merged.

Figure 9. Lucky Friday Mine. Headframe of the Silver Shaft (6,100 feet deep) in center; cement plant to the left.

In 1978, a failed copper mining venture at Lakeshore, Arizona, brought Hecla close to financial disaster. The company would not suffer for long, however. Within two years, the big rise in silver prices, sparked by market speculations in 1979-1980 by the Hunt brothers of Dallas, Texas, enabled the company to pay off its large debt and begin sinking the $40 million "Silver Shaft" at the Lucky Friday. The shaft is now 6,100 feet below the surface and the main production shaft at the mine. The 140-foot-high headframe stands in contrast to the old hoist works visible on the hill behind the mine offices.

Recently, Hecla has implemented a new mining method at the Friday called the Lucky Friday Underhand Longwall (LFUL). In the LFUL method, miners remove ore from the vein downwards (sink on the vein) rather than upwards (raising on the vein) as they do in the more standard cut-and-fill mining used in other mines in the district. As the miners sink on the vein they fill the resulting hole with concrete. The concrete is mixed at the surface in a new $2 million plant and then piped to the 5,000 foot level in the Silver Shaft and then to the working area. The LFUL method increases productivity, but more important it alleviates rock bursts. Rock bursts are explosive rock failures caused by mining-induced stress at the one-mile and more depth. Fatalities have been caused by rock bursts, and the LFUL method may decrease this danger.

The Lucky Friday Mine is in the Revett-St. Regis group of mines. It is also essentially a single vein like the Star-Morning, but it has been folded into a crude U-shape. As noted, the Friday may be one of the Link veins, like those in the Bunker Hill Mine, that have been transported eastward along the Osburn fault. The vein may have been folded as a result of drag along the fault. The Friday is noted for "blue rock" or disseminated sulfides in the Revett Quartzite. Disseminated sulfides are found in most of the Revett-St. Regis deposits but are more common in the Friday. The blue rock is the same material that was mined in the Snowstorm Mine at the next stop.

For Side Trip 6: Lucky Friday Mine to Snowstorm Dump—See Part 4: Side Trips.

Return to I-90

Part 4: Side Trips

Side Trip 1: Pinehurst to Page and the Page Mine

Turn off I-90 at Exit 45, Pinehurst. Turn left and go under I-90 and take the next right turn onto the old Pinehurst-Smelterville Road. It is 1.1 miles to the Page Road that turns south off the old Pinehurst-Smelterville Road. The Page Road makes a loop that includes the mine and the town. The road is in the Prichard Formation. The Osburn fault is near where a short spur road goes to the mine from the loop road. The fault separates Prichard Formation to the north from the St. Regis Formation to the south.

Side Trip 2: Up Big Creek to the Sunshine Mine

Set tripmeter to 0.0 (or record mileage) at the stop sign between the I-90 exit and the Big Creek Road. Turn right on Big Creek Road. At the fork in the road, keep to the left on Big Creek.

Mileage	Description
0.1	Railroad track and road to left to Polaris Peak. Go straight on Big Creek Road.
0.3	Sunshine Mine tailings impoundment.
1.1	Main office complex of Sunshine Mining Company. A Ken Lonn sculpture of a miner and his family is on the lawn. Inside is a display of mining paintings and a large model of the Sunshine Mine. The long building behind the office is the company's antimony plant where this metal is removed from tetrahedrite, the primary silver-bearing mineral in the silver belt mines. The building north of the antimony plant is the silver refinery where copper and silver are extracted from concentrates. With a mint in Coeur d'Alene, Sunshine is a vertically integrated operation as a miner, refiner, and minter of silver.
1.8	On the right is the Crescent Mine, owned by the Bunker Hill Company U.S. Ltd. The Crescent is the most westerly producing mine in the silver belt, a line of silver mines that includes the still-active Sunshine, Coeur, and Galena mines. All of the silver belt mines are on the overturned north limb of the Big Creek anticline, and all are in the Revett-St. Regis group. The Polaris fault is a major structure paralleling the silver belt.
2.2	On the left is the Sunshine Mine and the headframe of the Jewell shaft, the entrance to the mine. The Sunshine Mine was staked in 1884 by True and Dennis Blake, brothers from Maine, who had a homestead on Big Creek. The mine came into prominence in 1931 when a major strike of rich silver-bearing ore was made on the 1700 level. This level was also the intercept with the favorable Revett-St. Regis horizon. The "Shine" has produced over 350 million ounces of silver, probably the largest silver-production of any single silver mine in the world. By comparison, the fabled Comstock lode in Nevada produced about 200 million ounces of silver. Above the mine, north of the shaft, are the original dumps of the Blake brothers' early workings. The workings are in the Wallace Formation. South of the shaft is a magnificent exposure of the Big Creek anticline that is not visible from the mine site.

The core of the Big Creek anticline consists of thin interbeds of shales and flaggy sandstones of the St. Regis Formation. The south-dipping Big Creek fault on the southern flank of the anticline has a reverse throw. A thick sequence of Revett Quartzite is exposed along the road.

End of Side Trip 2. Return to I-90 and turn east to Wallace.

Side Trip 3: Up Lake Creek to the Galena Mine

Turn off I-90 at Exit 60. Turn east and follow the road that parallels I-90 to the south. Turn right at Lake Creek road, also known locally as Lake Gulch Road.

Mileage Description

0.0 Set tripmeter to 0.0 (or record mileage) at start of Lake Creek road.

The Belt sediments exposed along the road are overturned and dip steeply to the south. The lower Wallace Formation is exposed at the mouth of Lake Gulch.

0.7 The upper St. Regis Formation consists of a thinly bedded to laminated, light greenish yellow to greenish gray argillite unit. The unit is only about 150 feet thick and is porcellaneous in part. Most of the St. Regis is grayish red to reddish purple, thin interbeds of impure quartzite and darker argillite.

1.0 Buildings to the left are the Galena mill where ore is processed to concentrates by selective flotation.

1.1 Headframe of the Galena Mine. The trace of the Polaris fault crosses the valley just to the north of the Galena Mine shaft. The lower part of the Wallace Formation is south of the fault.

End of Side Trip 3. Return to start of Lake Creek road. Either return to I-90 or turn right and enter the city of Wallace.

Side Trip 4: Wallace up Ninemile Creek

Turn left onto 7th Street from Bank Street just before the Shoshone County Courthouse. Go past Hecla's previous corporate office building (last building on the right, formerly the Gyde-Taylor building before Hecla purchased it). Turn left onto Cedar Street. Go past the Star Mining Company's office. Turn right onto 6th street at the next intersection.

Mileage Description

0.0 Set tripmeter (or record mileage) at the intersection of 6th Street and the Burlington Northern Railroad tracks, just north of the old depot (built in 1902), now the Northern Pacific Depot Railroad Museum. Proceed straight ahead on U.S. Forest Service Road 456 that goes over Dobson Pass to Murray.

0.6 The Sierra Silver Mine is to the left. In 1987, over 12,000 visitors toured this popular attraction that features a visit to an underground mine and and an explanation of mining technology. It is one of the most popular tourist attractions in Idaho.

1.1 To the left is the historic Ninemile cemetery. James Callahan is buried here as are the three miners killed in the 1892 mining war at Gem.

1.6 Wallace Meat Company is across from the old townsite of McCarthy. The road crosses the Osburn fault, and the rocks change from Prichard Formation to Burke Formation. Another fault separates the Burke Formation from the Revett Quartzite.

2.3 Black Cloud Creek. Black Cloud fault

2.9 Dayrock Mine in St. Regis Formation. The Dayrock was originally one of Stratton's options. A novel way of financing mining operations for its time, the Strattons sold options on mining claims that could be redeemed for stock if the mine were successful. Their best property was the site of the Dayrock Mine. By the time the mine lived up to its promise, the Day family owned most of the options. The mine operated until 1977. The shaft of the Dayrock was sunk through the Dobson Pass fault.

3.3 East Fork/Ninemile Creek confluence.

Continue on the East Fork road to the south Gem stock. This dirt road goes to several mines including the Success (Granite), 16 to 1 (Rex), Interstate-Callahan, Tamarack, and Custer. The Interstate-Callahan and 16 to 1 are in the upper Prichard in the same horizon as the Pine Creek mines. The Interstate-Callahan was mined out just above the quartzite like the Sidney Mine, the largest producer on Pine Creek. All of these mines lie within the thermal aureole of the Gem stocks. The Success was used by some early geologists to show that the deposits were indeed hydrothermal and not derived from the sediments. This idea was discredited in later years as deep mining progressed. The Dobson Pass fault splits East Fork-Ninemile fork at the turnoff. Syenites of the Gem stock can be observed by stopping at the East Fork junction, walking due north, and climbing the hill to the old railroad grade. The syenites are exposed along the grade (see field trip by Schalck, this volume).

Continue straight ahead. Paved U.S. Forest Service Road 456 goes over Dobson Pass through the townsite of Delta and then on to Murray or Prichard on the North Fork of the Coeur d'Alene River.

Return to Wallace at the intersection of Bank Street (I-90) and 7th street.

Side Trip 5: Wallace to Burke

Mileage	Description
0.0	Set tripmeter to 0.0 (or record mileage) at the 7th and Bank Street intersection in Wallace at the County Courthouse. Go east past the Public Safety Building on the left. Turn north onto Idaho Highway 4 towards Burke.
0.2	Cross the railroad track at turnoff to Canyon Creek. Orange buildings just ahead are the machine shop of the Coeur d'Alene Hardware Company, a past producer of mining machinery.
0.5	Old Standard mill site on right. All that is left of the mill that processed ore from the Standard Mammoth Mine is one green building and a small brick building. The brick building housed the step-down transformer at the mill for electric power that was brought to the district in 1903. Wallace Formation is on both sides of road.
1.4	Woodlawn Park. The road crosses the Osburn fault, which separates the Wallace Formation on the south from the Prichard Formation on the north. On the right is the tailings impoundment for Hecla's Star-Morning mine.
2.8	Canyon Silver Mine (former Formosa Mine) on right.
3.6	Road sign, Gem. The Gem stock is exposed on the west side of the road. The Prichard Formation is on the east side. Gem was the site of the Hecla mill. It was difficult to build large mills in the narrow canyon in the same place as the mines, so the mills and mines were spread up and down the canyon taking advantage of every wide spot in the canyon bottom. The brick building (3.9 miles), which was Hecla's assay office for some years, spans Canyon Creek and is all that is left of the giant mill complex.
4.3	Enter Frisco. Frisco Mill site. The Panic of 1893 was the end of a calamitous fall in metal prices which caused the mine owners in the district to cut wages or close mines. Labor unions had existed in the district since Simeon Reed first cut miners' wages at the Bunker Hill in 1885. In 1892, the mine owners told the miners they were going to cut wages. The union men rebelled and went on strike. The owners countered by hiring non-union men from outside the district to run the mine. In addition, a Pinkerton detective named Charles Siringo, hired by the mine owners, infiltrated the union at Gem. Union activists discovered the deceit and this along with other provocations by the owners led to the dynamiting of the nearby Frisco mill by union supporters on July 11, 1892. This was the first so-called labor war in the Coeur d'Alene. Governor Willey requested federal troops to put down the lawlessness and restore order. An outgrowth of the conflict was the birth of a new labor union, the Western Federation of Miners. The organization would soon spread throughout the west. Many more violent confrontations would be the consequence in coming years.
4.5	Bridge over Canyon Creek. Tamarack Mine is visible ahead on left.
4.7	Black Bear townsite. Rocks are the Burke Formation. The north-south trending Frisco fault goes through here. The Tamarack Mine is across the creek ahead to the left, and the old Black Bear Mine is on the hillside to the right.
5.3	Bridge across Canyon Creek. The Standard fault cuts through here separating Burke Formation from Prichard Formation.
5.7	Road sign, Cornwall.
5.8	Concrete portal on the dump to the left of the creek is the Campbell No. 5 tunnel of the Standard-Mammoth (later the Green Hill-Cleveland) Mine.
5.9	Road sign, Mace. Mace is known for the 1910 snow slide that killed 17 people and destroyed several homes. The narrow canyon is subject to frequent snow slides brought down off the steep canyon walls by warm "Chinook" winds.
6.3	Road crosses Canyon Creek again. Old Hecla dump is on the right.
6.4	The town of Burke. Most of Burke is in the Prichard Formation but Burke Formation, is exposed near the end of the pavement. The Hecla Mine was the original property of the Hecla Mining Company formed in 1891. The headframe at Burke serviced the mine. Later, in 1921, Bunker Hill and Hecla formed the Sullivan Mining Company to mine the Star Mine, located at the west end of the Morning Mine, for rich zinc ore to feed the electrolytic zinc plant at Kellogg. The Star was accessed by a 8,900-foot-long tunnel from the 2000 level of the Hecla Mine. After Hecla was mined out, a new Star tunnel was

driven in 1953 from the canyon level to the Star workings.

Approaching the Star-Morning/Hecla surface plant, the date 1923 is visible on the end of the concrete flume that channels Canyon Creek. The date refers to the 1923 fire that destroyed the entire surface workings of the old Hecla plant and much of Burke and Gem. The surface plant was back in operation one year later with a new fire-proof structure that remains today. The dump on the left, across from the concrete building on the road, is the old Hecla extension.

7.0 End of the pavement. The timbers to the left are the site of the Sherman mill. The rough dirt road that turns to the left at the end of the pavement goes up Gorge Gulch and to the upper tunnels of the Hercules Mine. The Sherman Mine was owned by Day Mines. It was also the site of the long Hercules No. 5 tunnel that went from the mill site to the Hercules Mine. The brick building at the mouth of Gorge Gulch was the old Hercules office.

Just before the end of the pavement is where the Tiger Hotel and the Tiger and Poorman Mines were located. The hotel was unique because it was constructed over the railroad tracks. A tunnel was built through it to allow trains access to a loading platform for the Hercules Mine a little further up the canyon.

Burke was named for John Burke. The town was featured in Ripley's "Believe It or Not" for its unique layout. It was built in such a narrow canyon that merchants had to roll up their store awnings or lose them to passing trains. The town is also known as the birthplace of Lana Turner.

Return to Wallace, to the intersection of State Highway 4 (Canyon Creek road) and Bank Street (I-90).

Side Trip 6: Lucky Friday Mine to Snowstorm Mine Dump

Mileage Description

0.0 Lucky Friday Mine. Set tripmeter (or record mileage). Head east on the old road that parallels I-90.

0.8 Willow Creek drainage. Take the left fork in the road. Mullan Pass, Shoshone Park, and the Fish Hatchery are down the right fork.

1.4 Gentle Annie Gulch.

1.8 Snowstorm Mine dump, Silver Mountain shaft on Daisy Creek. Wallace Formation at dump site.

2.0 Road to Silver Mountain shaft headframe.

The Snowstorm Mine is different from other mines in the district. The ore at the Snowstorm is disseminated sulfides in the Revett Quartzite similar to that mined at Troy, Montana, and present in other known disseminated copper/silver deposits in the Revett in Montana and Idaho. As noted, the ore is also similar to the blue rock in other Revett-St. Regis mines in the district. The Snowstorm shipped ore to several smelters that needed the quartzite for flux.

In 1956, Hecla and Bunker Hill began a bold exploration venture by sinking a 2,000-foot shaft at the Silver Mountain project and then driving some 3,000 feet to the north through the Deadman syncline. Ground stability in the syncline was terrible. Long drill holes probed the area beneath the Snowstorm Mine, but economic mineralization was not discovered and the project was abandoned. Hecla began another exploration program in the Snowstorm in 1988 searching for low-grade silver-copper ore.

Return to I-90.

ACKNOWLEDGEMENTS

We would like to thank Victoria E. Mitchell (IGS) for drafting the figures for this paper. The work of the volume editors is also appreciated.

REFERENCES

Anderson, A. L. 1949, Monzonite intrusion and mineralization in the Coeur d'Alene district, Idaho: Economic Geology, v. 44, no. 3, p. 169-185

Bennett, E. H., 1984, A hypothesis concerning the genesis of orebodies in the Coeur d'Alene Mining District, Idaho: Idaho Geological Survey Technical Report, 84-7, 39p.

Bennett, E. H., and Ramesh Venkatakrishnan, 1982, A palinspastic reconstruction of the Coeur d'Alene Mining District based on ore deposits and structural data: Economic Geology, v. 77, no. 8, p. 1851-1866.

Bijak, M. K., and D. I. Norman, 1982, Mineralization of the Bunker Hill Mine, Coeur d'Alene district, Idaho, in light of fluid inclusion studies: Geological Society of America, Abstracts with Programs, v. 14. no. 7, p. 443.

Chevillon, V. C., 1977, Tectonically induced, Proterozoic soft sediment deformation at the Atlas property, Coeur d'Alene Mining District, Idaho: Geological Society of America, Abstracts with Programs, v. 9, no. 6, p. 716.

Conley, Cort, 1981, Idaho for the curious: Backeddy Books, Cambridge, Idaho, 704 p.

Fryklund, V. D., Jr., 1964, Ore deposits of the Coeur d'Alene district, Shoshone County, Idaho: U.S. Geological Survey Professional Paper 445, 103 p.

Gott, G. B., and J. B. Cathrall, 1980, Geochemical exploration studies in the Coeur d'Alene district, Idaho and Montana: U.S. Geological Survey Professional Paper 1116, 663 p.

Harrison, J. E., M. D. Kleinkopf, and J. D. Wells, 1980, Phanerozoic thrusting in Proterozoic Belt rocks, northwestern United States: Geology, v. 8, p. 407-411

Harrison, J. E., A. B. Griggs, and J. D. Wells, 1974, Tectonic features of the Precambrian Belt basin and their influence on post-Belt structures: U. S. Geological Survey Professional Paper 86, 15 p.

Hershey, O. H., 1916, Origin and distribution of ore in the Coeur d'Alene (private publication): Mining and Scientific Press, San Francisco, 32 p.

Hobbs, S. W., A. B. Griggs, R. E. Wallace, and A. B. Campbell, 1965, Geology of the Coeur d'Alene district, Shoshone County, Idaho: U.S. Geological Survey Professional Paper 478, 139 p.

Hobbs, S. W., and V. C. Fryklund, 1968, The Coeur d'Alene district Idaho, *in* Ore Deposits of the United States, V. 2 (Graton-Sales Volume): American Institute of Mining Engineers, New York, p. 1417-1435.

Juras, D. S., 1982, Structure of the Bunker Hill Mine, Kellogg, Idaho, *in* R. R. Reid and G. A. Williams, editors, Society of Economic Geologists' Coeur d'Alene Field Conference, Idaho — 1977: Idaho Bureau of Mines and Geology Bulletin 24, p. 31-34.

Landis, G. P., D. L. Leach, and A. H. Hofstra, 1983, Silver-base metal mineralization as a product of regional metamorphism, Coeur d'Alene district, Shoshone County, Idaho: Concepts of genesis (Abstract), *in* S. W. Hobbs, editor, Belt Symposium II, Program and Abstracts: Montana Bureau of Mines and Geology Special Publication 90, p. 68.

Landis, G. P., and D. L. Leach, 1983, Silver-base metal mineralization as a product of metamorphism, Coeur d'Alene district, Shoshone County, Idaho: Geological Society of America, Abstracts with Programs, v. 15, no. 5, p. 327.

Leach, D. L., and A. H. Hofstra, 1983, Fluid inclusion studies in the Coeur d'Alene district, Idaho: Geological Society of America, Abstracts with Programs, v. 15, no. 5, p. 327

Leach D. L., G. P. Landis, and A. H. Hofstra, 1988, Metamorphic origin of the Coeur d'Alene base- and precious-metal veins in the Belt basin, Idaho and Montana: Geology, v. 16, p. 122-125

Ramalingaswamy, V. M. and E. S. Cheney, 1982, Stratiform mineralization and origin of some of the vein deposits, Bunker Hill Mine, Coeur d'Alene district, Idaho, *in* R. R. Reid and G. A. Williams, editors, Society of Econmoic Geologists' Coeur d'Alene Field Conference, Idaho — 1977: Idaho Bureau of Mines and Geology Bulletin 24, p. 35-43.

Ransome, F. L., and F. C. Calkins, 1908, The geology and ore deposits of the Coeur d'Alene district, Idaho: U.S. Geological Survey Professional Paper 62, 203 p.

Reynolds, M. W., and M. D. Kleinkopf, 1977, The Lewis and Clark line, Montana-Idaho: A major intraplate tectonic boundary: Geological Society of America, Abstracts with Programs, v. 9, p. 1140-1141.

Simos, J. G., and K. J. Droste, 1961, A study of the northeastern part of the Coeur d'Alene district: Private report for Hecla Mining Company.

Sorenson, A. H., 1968, Genesis, mode of origin, and age of the Coeur d'Alene ore deposits, Idaho: private publication, 89 p.

Springer, D. C., 1987, Coeur d'Alene mining production history: private publication, 1 p.

Umpleby, J. B., 1917, Genesis of the Success zinc-lead deposit, Coeur d'Alene district, Idaho: Economic Geology, v. 12, no. 2, p. 138-153.

Umpleby, J. B., and E. L. Jones, Jr., 1923, Geology and ore deposits of Shoshone County, Idaho: U.S. Geological Survey Bulletin 732, 156 p.

Webster, T. A., 1981, Faulting and slumping during depostion of the Precambrian Prichard Formation near Quinns, Montana: A model for sulfide deposition: University of Montana M.S. Thesis, 71 p.

White, B. G., in press, Superposed map-scale folds and subsequent veins unrelated to Osburn strike-slip fault, Coeur d'Alene Mining District, Shoshone County, Idaho: Geological Society of America, Abstracts with Programs.

White, W. H., 1959, Cordilleran tectonics in British Columbia: American Association of Petroleum Geologists Bulletin, v. 43, p. 60-100.

Zartman, R. E., and J. S. Stacey, 1971, Lead isotopes and mineralization age in Belt Supergroup rocks, northwestern Montana and northern Idaho: Economic Geology, v. 66, no. 6, p. 849-860.